THE METAVERSE OF CONSCIOUSNESS

"This book investigates the true nature of reality from a psychonaut's point of view. Through awareness of perceiving the Whole, we can envision and participate in evolution of human consciousness, rising above apparent divisions and polarities. Shelli Renée Joye invites us to experience the richness of our living Cosmos and the dawning of a new awareness of what it is to be truly human, sharing ideas of Teilhard de Chardin, David Bohm, Carl Jung, G. I. Gurdjieff, and more."

CYNTHIA SUE LARSON, PH.D., AUTHOR OF
QUANTUM JUMPS AND *REALITY SHIFTS*

"In *The Metaverse of Consciousness*, Joye blends hard science theory, new paradigm cosmology, consciousness study, and brain science with autobiography, psychedelic vision, and self-identity to address the very question of what personal experience is, where it happens, and how we can expand it. Her style is engaging, the material interesting, her scope wide-ranging, and her visions ultimately hopeful. The universe is larger and deeper than we think. Go see!"

TOBY JOHNSON, PH.D., AUTHOR OF *FINDING YOUR OWN TRUE MYTH*

"Joye brings together her personal search for the structures of consciousness, the concepts of Jung's psyche and psychoid structures, Teihard's hyperphysics and evolution of consciousness, and Pribram and Bohm's holoflux of consciousness. She identifies commonalities of mystical insight through the ages with current scientific investigation with a rigor that weaves together the historical concepts describing consciousness with contemporary scientific findings. She pushes the envelope to provide a plausible explanation of the source of consciousness in a quantum world that interacts with the spacetime continuum in which we experience our existence. Her style of writing is easy to read, and the concepts she discusses will push any imagination toward a new understanding of the world in which we live."

JOHN ALAN TAYLOR, PH.D., DEAN AT ORANGE COAST COLLEGE

"This book offers hope and direction to those who seek guidance but have become discouraged by the heavy hand of scientific materialism. Weaving an integrated tapestry of ideas from the core visionary works of Carl Jung, David Bohm, Teilhard de Chardin, and Gustav Fechner, the book offers credible answers to our most fundamental questions while constantly inviting the reader to participate in a direct introspective exploration of our vast unfolding noospheric metaverse. Opening with a rich, detailed exposition of the hidden dimensions of reality, supported by recent data from string theory and particle physics, the book offers a new way of seeing the world as a metaverse. Clearly written with love, great attention to detail, and a wealth of visual diagrams. Highly recommend."

MICHAEL PRYZDIA, PH.D., AUTHOR OF *A MATURE MYSTIC*

"Joye takes us by the hand and whispers tales of mind explorers who dared to work at the edges of consciousness. Walking the fine line between inspired discipline and intellectual kindness, *The Metaverse of Consciousness* arrives at a prefatory synthesis that integrates deep experience, revolutionary science, and subtle metaphysics. The conundrum of world and mind meets halfway in us humans. This book is a timely contribution to the healing of our civilization's self-inflicted wounds."

ÀLEX GÓMEZ-MARÍN, PH.D., ASSOCIATE PROFESSOR AT INSTITUTO DE NEUROCIENCIAS AND DIRECTOR AT PARI CENTER

"Joye has brought the Jungian system into a clearer focus than I have encountered before. The same is true for her chapter on Teilhard de Chardin. The illustrations in the book highlight the overall map of the theories being expounded conceptually, making this book a great resource for graduate students and those teaching on the subject. A unique aspect of this book is that these theories were not merely abstract to the scientists involved, but Joye relates how each of them directly experienced these fundamental realities."

DANIEL KEALEY, PH.D., PROFESSOR EMERITUS OF THE DEPARTMENT OF PHILOSOPHY AND RELIGIOUS STUDIES, SISKIYOUS UNIVERSITY, AND AUTHOR OF *REVISIONING ENVIRONMENTAL ETHICS*

THE **METAVERSE** OF **CONSCIOUSNESS**

Mapping the
Multiple Dimensions of Reality

SHELLI RENÉE JOYE, Ph.D.

Inner Traditions
Rochester, Vermont

Inner Traditions
One Park Street
Rochester, Vermont 05767
www.InnerTraditions.com

Copyright © 2025 by Shelli Renée Joye, Ph.D.

Technical material in chapters 6 and 7 is based on the author's 2016 Ph.D. dissertation, "The Pribram-Bohm Holoflux Theory of Consciousness: An Integral Interpretation of the Theories of Karl Pribram, David Bohm, and Pierre Teilhard De Chardin," and also appeared in her book published by Inner Traditions in 2018, *The Electromagnetic Brain: EM Field Theories on the Nature of Consciousness.*

All rights reserved. No part of this book may be reproduced or utilized in any form or by any means, electronic or mechanical, including photocopying, recording, or any information storage and retrieval system, without permission in writing from the publisher. No part of this book may be used or reproduced to train artificial intelligence technologies or systems.

Cataloging-in-Publication Data for this title is available from the Library of Congress

ISBN 979-8-88850-012-5 (print)
ISBN 979-8-88850-013-2 (ebook)

Printed and bound in India at Replika Press Pvt. Ltd.

10 9 8 7 6 5 4 3 2 1

Text design and layout by Kenleigh Manseau
This book was typeset in Garamond Premier Pro with Refrigerator Deluxe and Mr Eaves Mod OT used as display type faces

All images are either produced and copyrighted by the author, figures that exist in the public domain, or images released into the creative commons. Figs. 1.1, 1.2, 1.3, and 1.5 CC BY-SA 3.0; Fig. 6.1 CC BY 2.5; Fig. 12.2 CC BY-SA 2.1 JP.

To send correspondence to the author of this book, mail a first-class letter to the author c/o Inner Traditions • Bear & Company, One Park Street, Rochester, VT 05767, and we will forward the communication, or contact the author directly at **shellijoye.net**.

Scan the QR code and save 25% at InnerTraditions.com. Browse over 2,000 titles on spirituality, the occult, ancient mysteries, new science, holistic health, and natural medicine.

*To the true treasures of my life, my children:
my son, Jason Kenneth Joye, who earned his M.S. in computer
engineering and is now a senior managing software engineer in
Santa Clara, California; and my daughter, Alyssa Maria Joye,
who, after completing her graduate studies at the
London School of Economics, found her calling as
a clinical psychologist with the NHS in London.*

*Also, to my dear friend, Allan Combs, President Emeritus of the
Society for Consciousness Studies, who guided me through the
dissertation process with wisdom and patience; and to my close
friend and colleague, Michael Pryzdia, a professor of
philosophy and consciousness studies at Arizona State University,
for his unwavering support and insightful discussions.*

The day science begins to study non-physical phenomena, it will make more progress in one decade than in all the previous centuries of its existence
                Nikola Tesla, "The Law of Reflection"

The physical world is the way the mental world concieves of and visualizes itself.
                Toby Johnson, Ph.D.

Contents

Foreword   ix
    By Allan Combs, Ph.D.

Introduction   1
    Reader's Guide   9
    Key Terms and Definitions   16

PART 1
THE HIDDEN DIMENSIONS OF CONSCIOUSNESS

1   What Are the Hidden Dimensions?   20
2   Universe or Metaverse?   32
3   How I Came to Explore the Hidden Dimensions   48

PART 2
THE PSYCHOPHYSICAL METAVERSE

4   Fechner's Psychophysics of the Human Soul   80
5   Jung's Spectrum of the Psyche   90
6   Teilhard's Sense of Collective Consciousness   123

PART 3

HYPERPHYSICS OF THE NOOSPHERE

7	Hyperphysics: Speculative Physics beyond Material Science	150
8	The Energetic Noosphere	171
9	The Superposition of Consciousness	200

PART 4

SUB-QUANTUM CONSCIOUSNESS

10	The Pribram-Bohm Holoflux Theory	212
11	The Digital Universe	236

PART 5

CONCLUSIONS

12	Exploring the Hidden Dimensions of the Mind	264

• • •

Notes	278
Bibliography	291
Index	300

Foreword

Allan Combs, Ph.D.

In this book are stories of several remarkable explorers of inner space, *psychonauts*, who returned from journeys into deep consciousness to report their findings to the rest of the world. They might be described as scientist-mystics, owing to their understanding and integration of science and mysticism. It is also the story of efforts to understand the physics and metaphysics of the vast cosmos in which we live, now known as the *metaverse*. But first, let us speak of reality itself, which seems to get bigger every day; a fact that can be unsettling.

Such news about reality has been coming for a long time. The great sea navigators of the Renaissance, such as Ferdinand Magellan and Vasco da Gama of Portugal, and Christopher Columbus of Italy, brought home disquieting stories of a vastly bigger Earth than people at that time even suspected. An Earth in which all of Europe was only a small part.

Meanwhile, astronomers of that period such as Copernicus, Kepler, and Galileo were publishing radical ideas about the nature of the sky and the celestial cosmos—and worse yet, backing them up with observations made with newly invented telescopes, and with sophisticated mathematics. Giordano Bruno, who espoused the idea of infinite space and the location of the sun at the center of the solar system, was burned at the stake for his beliefs. So you see how deeply disturbing these ideas of a greater cosmos were to the people of those times. Indeed, there was a

Fig. F.1. The fall of Phaéton, 1588.

growing sense of vertigo, or anxiety, that permeated the art and thought of the era.

This situation is artistically represented by a Dutch engraving of Phaéton, son of the sun god, Helios, falling from the sky, unable to keep his balance and control the horses that drove his father's chariot of the sun across the sky each day.

Now, in the early decades of the twenty-first century, we are again faced with a cosmos that grows larger by the day. And again, we are confronted by naysayers who have replaced the bishops and cardinals of the Renaissance, and again cling to beliefs crafted long ago; ideas that were progressive in their time, during the European Enlightenment, but now seem to best represent Democritus's famous dictum, "All that exists is atoms and void." But that's not the half of it. These psychonauts have opened pathways into deep consciousness that are now being followed by many others.

It is no secret to those interested in such matters, including present readers, that individuals in cultures around the world, and across history, have reported many similar experiences of realities beyond the everyday waking world. These include a wide range of dreams, some of which seem to take place in remarkably authentic otherworldly settings. They also include shamanic journeying into upper and lower realms, similar globally in many different cultural milieus. Reports of near-death experiences are widespread these days as well, and have been validated in many scientific reports. Even postmortem reports of communication with deceased persons, common and studied in the Victorian Age by scientists as well respected as William James, have now returned, bringing scientific scrutiny with it. Perhaps also worth mentioning, every major investigator of UFOs (now termed UAPs or "unidentified aerial phenomena") has given up on explaining them in material terms. This includes the late J. Allen Hynek, professor of astronomy and early ufologist; and Jacques Vallée, leading contemporary ufologist, who has suggested that UAPs may be a manifestation of trans-dimensional or even transhistorical intelligence.

All the above point to a larger and layered metaverse, vastly greater than imagined even fifty years ago. It also points to the possibility of a future science that integrates human experiential reality, that is, human consciousness, with a future physical science, so much as the latter has retained any meaning in this era of quantum physics. The book you hold in your hands, or view on your computer screen, is a giant step in this direction. Please enjoy the fascinating story that unfolds.

ALLAN COMBS, PH.D., is Emeritus Professor of Consciousness Studies at the California Institute of Integral Studies and the University of North Carolina-Asheville. He is author of over 250 publications on consciousness and the brain, and is the founder and past president of the international Society for Consciousness Studies, as well as the recent recipient of their Lifetime Achievement Award. He is also cofounder of the Society for Chaos Theory in Psychology and the Life

Sciences. Combs is editor of the journals *Consciousness: Ideas and Research for the Twenty-First Century* and the *Journal of Conscious Evolution*. He was also the winner of the National Teaching Award of the Association of Graduate Liberal Studies. Combs's most widely read books include *Consciousness Explained Better: Towards an Integral Understanding of the Multifaceted Nature of Consciousness*, and *The Radiance of Being: Understanding the Grand Integral Vision, Living the Integral Life*.

Introduction

In this age of confusion, disinformation, and conflicting belief systems, where can we turn to find answers to our most basic questions?

- "Where did we come from?"
- "Why are we here?"
- "Where are we going?" and
- "What can we do to move more safely into the future?"

These have always been the basic questions, the answers to which can nourish our identity and guide us, both as individuals and as groups. But today these questions have become even more significant because, in a very real sense, we have lost our way. Even nature seems to have turned against us. We can no longer be sure of the direction that our lives are taking, and a familiar world we have always taken for granted is quickly slipping away, lost and transformed at an accelerating pace. If only we knew where to look for a map or a source we might identify to answer such questions, encourage us, and guide us safely into the future!

It is easy to become discouraged and paralyzed by the endless barrage of bad news. "Doom-scrolling" has become not just a buzzword, but an obsession. Millions are losing faith not only in religious traditions, but in governments, in economic systems, and in the very ways we have lived our lives; a growing wave of social media is clearly beginning to doubt our ability to survive as a species. Many are beginning to realize that a radical transformation is required, an immediate shift in the way we live, work, eat, and interact, individually and collectively, if we are to continue to survive as a civilized species.

For many, religions offer ways to deal with the crisis. While they give a sense of continuity, stability, and a focus on inner tranquility and peace, their traditional teachings seldom provide answers to the current conflagration of social and environmental emergencies. Traditional religions urge us to pray for peace and patience, but if we want more direct, specific guidance and answers, where do we even begin to look? Our contemporary governments, religions, cultures, and educational systems all seem inadequate to the task.

Yet might there be other sources of knowledge and guidance to help us to deal with the complexities of accelerating environmental collapse, rising fascism, weakening of faith, and loss of hope?

Scientists tell us that the universe is mechanical, a "dead thing." They try to convince us that the universe runs like a machine, like some mechanical clock that has evolved only through random chance. In contrast, at the other end of the spectrum, multitudes of priests, artists, and mystics describe with great sincerity how they have been able to experience direct cognitive contact with invisible conscious entities, how they have come to experience "a flow" of awareness among beings that appear to inhabit a vast multidimensional universe. If this is true, the universe is definitely not a "dead thing"; on the contrary, generations of seers, mystics, and psychonauts* have insisted that the universe is full of sentient, conscious entities, or psychoids† at every scale, interacting and evolving in a wild dynamic process throughout the ocean of mysterious energy forms.

---

*Psychonauts: From the Greek ψυχή or *psychē* ("soul," "spirit," or "mind") and ναύτης or *naútēs* ("sailor" or "navigator"), thus a "sailor of the soul"; one who explores the oceans of consciousness that are normally hidden, regions of awareness traditionally accessed through means of prayer, fasting, meditation, entheogenic drugs, and innumerable varieties of psychophysical exercises (e.g., yoga, tantra, mantra, chant, drumming, etc.).

†Psychoids: Entities of the collective conscious/unconscious, which, as Carl Jung wrote "cannot be easily perceived or represented, are transcendental, and for that reason I have called them 'psychoids'," found in *The Structure and Dynamics of the Psyche*, vol. 8, *Collected Works*, paragraph 840. These psychoids can be even more fundamental to consciousness than the archetypes, which bridge the gap between human consciousness and the wide variety of psychoidal processes/entities.

To which of the two groups should we turn? Do we assume that all material science assumptions are correct beyond a doubt, that the material universe is, indeed, a dead, mechanical thing? The many technological wonders brought to us by material science (e.g., electricity, computers, vehicles, space stations, the internet, etc.) tend to make us receptive to the dogma often repeated by material science that "other than life on Earth, the universe is dead."

Clearly many of the successes of technology have been double-edged swords. Two centuries of burning hydrocarbons to power global transportation systems (steamships, automobiles, trains, airplanes) has already radically disrupted the planet's environment and weather patterns. The ubiquitous effect of social media, powered by digital devices and networks has, rather than bringing people closer together as initially envisioned, so isolated multitudes that they lose all sense of social cohesion, and a growing number of individuals become either lone actors or passionate members of inflammatory subgroups, fostering fragmentation, social chaos, and even violence. But perhaps most destructive of all, though largely unrecognized, has been the stifling effect that the scientific materialist paradigm has had on human efforts to advance the search for answers to what it means to be alive, to be conscious in a universe of space, time, and sensation.

Yet the religious, mystical view of the universe is equally problematical. It is true that arguments for a living, loving universe are deeply embedded within our cultural traditions over millennia, long before science was even a concept. However, religious tradition and even the language of mythology is quickly discounted by many modern readers, who have become now more open to the languages of science than the messages handed down through their own ancestral traditions.

It appears that *both* science and religion fail to address our major contemporary issues, both as individuals and collectively as families, nations, and humankind. Where then might we find sources of guidance to restore new meaning, direction, and hope to our lives here and now? Many scientifically trained thinkers down through the ages (including Isaac Newton, Rudolph Steiner, and even Albert Einstein) have insisted that there does exist a rich domain of metaphysical knowledge, guidance, and creativity that is beyond space and time, yet experientially accessible to human beings.

While religious traditions and schools of mysticism have empirically developed techniques to communicate with dimensions of being and consciousness that are normally hidden to human awareness. Part of the difficulty lies in the fact that even contemplatives often remain enmeshed in images and metaphors and modes of awareness that are based in space and time dimensions. The key to progress is to learn how to cultivate trans-spatial, trans-temporal modes of awareness by practicing to develop the skills to go beyond our normal ways of perceiving and thinking in space-time. Being able to perceive "the Voice of the Silence"* (the transcendental silence of the dimensions beyond space and time) allows a new way of "seeing" to arise, as one begins to perceive regions of ontological reality that project into time and space but that are primarily active within the hidden dimensions.

Virtually all of our modern material science, psychology included, has failed to pursue direct firsthand exploration of the inner spaces of consciousness. Now and in the twentieth century, only a very few of the scientifically trained were moved to break with the mainstream sufficiently to explore inner space through their own direct awareness (e.g., William James, Carl Jung, Teilhard de Chardin, John Lilly, William Tiller, Timothy Leary, Richard Alpert, among others).

This book begins with the premise that, unlike the dead, static, mechanical universe espoused by materialist thinkers, the universe is indeed full of life and consciousness in multiple dimensions. Throughout millennia we find oral and written accounts of human communication with other conscious entities, frequently referred to as angels, spirits, gods. There are many accounts of communication even with recognizable spirits of the dead. These sources are said to have been able to offer guidance and knowledge to the mystics, saints, and shamans who have been able to establish contact with these entities. It is hoped that the integrated discussion of "inner space" presented in this book, supported by contemporary physics, may assist humans to navigate these perilous times.

Neither technical innovations nor political action seem to be able to solve our problems, but perhaps a major awakening in understanding our role within this universe can lead us forward beyond the paralysis into

---

*Madame Blavatsky, *The Voice of the Silence* (1889).

which we have fallen, individually and collectively. In order to make sense of the currently mounting maelstrom of global and individual crises, trained psychonauts might investigate and adapt methods (e.g., contemplation, drugs, physical and mental exercises) to dissolve the normal boundaries of isolated individual awareness in order to establish direct communication with nonhuman psychoidal entities. Like fractals, psychoid consciousness agents exist in many forms, networking at all dimensional levels of inner and outer space.

Critical to this effort is the integration of scientific knowledge of *inner space* through the development and support of science-trained explorers of consciousness, contemporary psychonauts. An integrated understanding of science and consciousness, physics and metaphysics, should be a requisite for the new generation of scientist-psychonauts embarking on this quest to save our planet and our species.

One widespread misconception, based upon our direct, lifelong sensory experiences as human beings, is that we live solely within the dimensions of *space* and *time*. However recent developments in cosmological physics reveal evidence that there exist multiple regions of reality that appear to be *beyond* our familiar space and time. All dimensional regions of the larger reality may be fully interconnected and perceivable from the space and time dimensions in which we experience our daily lives.

The Universe and Spacetime

The word "universe" has been defined in physics as "everything within space and time."[1] For centuries scientists have assumed that there can be nothing outside of the universe of space and time, and that the two dimensions are inextricably connected. In 1909, to indicate how space and time (and everything within space and time) are interconnected, a new word came into use in scientific papers, *spacetime*.

Spacetime was formally introduced in 1908 by Hermann Minkowski during a lecture in Vienna. Minkowski had been a favorite professor of Einstein in his final student years, and was lecturing on the special theory of relativity his famous student had published three years earlier. Einstein's theory of relativity described space and time as inseparable, interwoven entities. During his lecture, Minkowski began using the new term *spacetime* as shorthand for Einstein's model of an integral, four-dimensional space and time continuum. The term, later referred to as "Minkowski spacetime," quickly spread to become widely accepted in publications by physicists and astrophysicists.[2] Scientists, until recently, have operated under the assumption that four-dimensional spacetime is "all that there is." This has been one of the few facts that all physicists agreed upon.

But physicists were caught off guard when data from recent high-energy particle experiments pointed to the existence of not one, but *seven* additional dimensions beyond time and space. The existence of these dimensions had been predicted by a mathematical model called superstring theory, a new field that had begun in the 1960s in the effort to develop a quantum theory of gravity. Faced with the evidence from particle experiments and supported by string theory, a number of scientists began to consider that multiple dimensions might actually exist, that there could very well exist an ontological reality beyond spacetime. One of these, Donald Hoffman, a project leader at the MIT Artificial Intelligence Laboratory, went so far as to state the following:

Spacetime was a wonderful framework for centuries but it's over now, it's time to get a new and deeper framework. We can use mathematics and go entirely outside of spacetime. The mathematics is network theory.[3]

Consciousness Agents in Neural Networks

Hoffman argues that while the primary regions of consciousness are vast and many, they mostly lie beyond, not within, the dimensions of spacetime. Currently a professor of cognitive science at the University of California, Irvine, he supervises a team of mathematicians and physicists engaged in developing a detailed mathematical map of consciousness and neural networks. The model promotes the idea of "agential consciousness," a focus upon relationships in consciousness, under the assumption that conscious entities work through networks of "conscious agents,"— much like modern AI computer architecture works through software neural networks.

Hoffman describes our psyches within spacetime as "virtual headsets" used by conscious agents. While neural networks exist within spacetime, conscious agents and their networks of consciousness have physical substrates that exist outside of spacetime. Although these agents or entities operate entirely outside of spacetime, they use spacetime as a communication interface. Hoffman suggests that the mathematics his team is developing supports the idea that our psyche, our "spacetime headset," can then help our separate spacetime psyches to create art (such as music, painting, dance, drama, or comedy), engage in complex games involving individuals, tribes, and nations, and explore innumerable forms of experimentation. Agential consciousness is seen as a fundamental unit of consciousness, a building block upon which hierarchies of networked conscious agents can arise to different levels, above those that coalesce beneath them, similar to levels on a pyramid (where the current "I" level is at the top of the pyramid). Multiple links can be established among pyramidical structures, creating even wider conscious networks. This allows for the evolution of innumerable unique "conscious agent" configurations to inhabit every dimension.

While most contemporary brain scientists continue to assume that consciousness is a recent phenomenon that is somehow caused by the firing of neurons in the brain, Hoffman argues that this assumption is highly overrated; he does not believe that neurons create consciousness, but that consciousness creates neurons. In fact, his mathematical model reveals networks of consciousness that operate out of the hidden dimensions to project what we call physical neurons into spacetime.

In his presentations, Hoffman encourages fellow scientists in all fields to explore the potential impact of these additional dimensions on their own research. Through his mathematical analysis of superstring theory, Hoffman reaches a conclusion that aligns with Bohm's earlier claim: the fundamental seat of consciousness resides *beyond* spacetime, primarily within the hidden dimensions that remain largely unexplored.

The spacetime headset, as Hoffman describes it, employs a relatively simple data compression scheme that conscious agents use to interpret their interactions with others. Interestingly, these conscious entities can also merge; for instance, simpler agents can combine in unique ways to form more complex agents. Hoffman suggests that death is not truly an end; rather, it is akin to removing the spacetime headset.

A Metaverse of Multiple Dimensions

To express this expanded notion of multiple dimensions, a new word, *metaverse*, has recently come into use. Used throughout this book, the term *metaverse* incorporates our familiar four-dimensional spacetime universe, plus the additional seven dimensions that have been discovered through high-energy experiments in quantum particle physics such as the discovery of the Higgs Boson Particle.

Yet the general public has not yet been made aware of these discoveries. Instead, the contemporary understanding of the word *metaverse* has been co-opted by software engineers, gamers, and social media platforms such as Meta (formerly known as Facebook). To this community, the concept of the metaverse implies various virtual reality "worlds" that a user can experience directly with a personal avatar, usually by means of Virtual Reality (VR) headsets.

By encouraging the exploration of consciousness however, these virtual reality worlds may be a blessing! These immersive platforms invite hundreds of thousands of individuals who enter these software worlds to experience what it is like to be conscious *outside of* or *beyond* our normal "spacetime world," in some separate reality. Persons using virtual reality headset devices (for example the Oculus headset) for gaming might eventually grow to find the experience not merely as entertainment, but as a stepping stone to exploring the previously unperceived worlds of consciousness that Hoffman and Bohm believe to exist beyond our normal awareness in spacetime. Gamers may find themselves becoming contemporary psychonauts, exploring the wider realms of consciousness in the wider metaverse, particularly as the former strict legal barriers to experimenting with consciousness assisted by ayahuasca and psilocybin and other formerly illegal entheogens have been falling away around the world.

Contemporary human psychonauts may be seen as following in the footsteps of the great explorers of the eighteenth and nineteenth century. Where those explorers opened the realms of our physical plane of existence by entering uncharted regions of awareness, these modern psychonauts will be instrumental in opening the uncharted regions of the newly discovered dimensions. Thus, the gaming and social media metaverse may well be the ideal training ground for individuals who will join in the exploration of these hidden dimensions, both by putting into practice the skills and insights garnered through their online and virtual interactions, and through the practice of various techniques used by mystics, shamans, and yogis for millennia, and by modern day psychonauts in making their first journeys into this wonderous new horizons of exploration through the use and application of:

- prayer
- meditation
- entheogenic experiences
- transcranial electromagnetic stimulation
- the new field of direct access via implantation of computer interfaces implanted within the brain.

READER'S GUIDE

The material in this book offers psychonauts (and those interested in consciousness) an integrated, contemporary, multidimensional view of consciousness in the metaverse. The book has been written using an integral approach* to offer readers of diverse backgrounds, education, and experience multiple ways of viewing the material so as to make the conclusions more readily understandable. Figure I.1 (p. 10) shows eight major perspectives that are used in developing an understanding of consciousness in the metaverse. Using these multiple perspectives has helped me, and hopefully will help the reader, to knit together more clearly an understanding of the role of nature and role of consciousness throughout the dimensions of the metaverse.

With these eight approaches in mind, I have developed this book as a "map" to better understand consciousness and the metaverse. The chapters have been written to be accessible whether one is a scientist or simply a curious reader. In all cases I have worked to make the science clear through the use of numerous figures and diagrams, while minimizing the use of equations, other than the all-important Fourier transform equation, discussed in chapter 9.

---

*An integral, or integrative, approach to knowledge is one that brings together multiple disparate perspectives, while seeking to integrate their most essential aspects into a single coherent, comprehensive whole that can then become the focus of discussion.

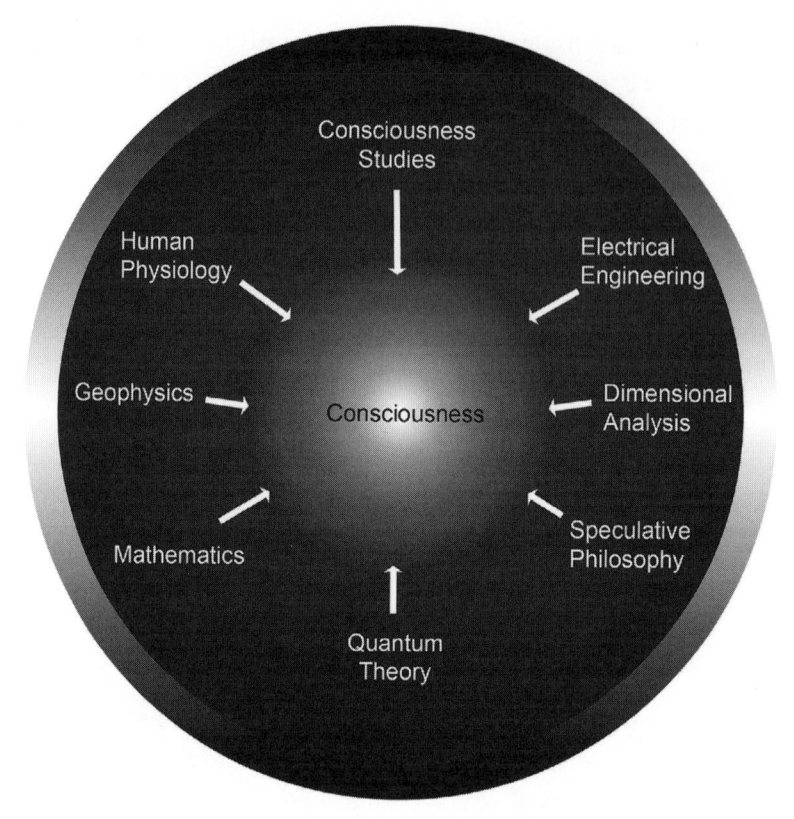

Fig. I.1. Approaches to understanding consciousness taken in this book.

What follows are short descriptions of chapter contents and themes. Note that it is not imperative that chapters be read sequentially. Each offers a fairly unique layer of information that can be absorbed by the reader in evolving his or her own understanding of consciousness and the metaverse.

Part 1:
The Hidden Dimensions of Consciousness

Overview: A breakdown of the dimensions both within and outside of our perception, and the scientific theories that examine and hypothesize the existence and interconnectivity of these dimensions. This section also touches on my inspirations to study these topics, and how we can gain better access and connectivity to the greater metaverse.

Chapter 1: What Are the Hidden Dimensions?

Summary: We begin by discussing the ontological reality of a wider universe (the metaverse) in the seven hidden dimensions discovered through recent high-energy experiments in quantum physics. All of these dimensional regions of the larger metaverse may be fully interconnected and perceivable from the space and time dimensions in which we experience our daily lives.

Key concepts and themes: Experimental proof of seven hidden dimensions through data gathered from particle colliders such as the Large Hadron Collider in Geneva; the potential measurement of these dimensions through theories such as string theory and M-theory.

Chapter 2: Universe or Metaverse?

Summary: This chapter implies that the word *universe* is an outmoded term to describe our experience of existence. A new word, *metaverse*, implies a much larger reality than that perceived by our ancestors, one that includes at least seven unique dimensions beyond the four of time and space. This chapter also examines living patterns of consciousness known as archetypal *psychoids,* that utilize symbolism to connect human consciousness with the metaverse.

Key concepts and themes: Morphic resonance and collective memory; the frequency spectrum; electromagnetic field theory as it relates to consciousness; the need to reconnect with psychoids through psychophysical exercises, visualizations, and entheogens.

Chapter 3: How I Came to Explore the Hidden Dimensions

Summary: This chapter serves as an introduction to the personalization of dimensional exploration and the importance of having multiple perspectives with which to approach an understanding of consciousness and our role in the metaverse. It contains autobiographical insight into my discovery of the hidden dimensions, through meditation, psychedelics, and spiritual exploration.

Key concepts and themes: How the author's approach to an understanding and a mapping of the wider metaverse was fostered by the direct experiences of psychedelic entheogens as well as a rigorous academic education in physics, metaphysics, and Asian philosophies.

Part 2:
The Psychophysical Metaverse

Overview: Early detailed approaches to using science to understand consciousness are discussed here in the context of the life work of three major scientists in the nineteenth and early twentieth centuries: Gustav Fechner, Carl Jung, and Pierre Teilhard de Chardin.

Chapter 4: Fechner's Psychophysics of the Human Soul

Summary: This chapter focuses on Fechner's life's work as a metaphysician, using mathematics and the tools of science to discover the physics of the human psyche. Fechner studied the human brain to facilitate the exploration of consciousness and possibility of communing with conscious entities.

Key concepts and themes: The concept that the entire material universe is conscious in a wide range of wavelengths of energy; Fechner's theory of the likely existence of two personalities (two minds, in one brain.)

Chapter 5: Jung's Spectrum of the Psyche

Summary: This chapter lays out a map of Jung's exploration of what he termed *psychoids*, quasi-independent entities of consciousness that fill the universe in every dimension. Jung worked to develop an integrated model of the psyche, an expanded map of the unconscious (both the individual unconscious and the collective unconscious), with an emphasis upon the effects of these collective unconscious psychoids upon individuation and wholeness in the individual.

Key concepts and themes: The introduction of archetypes and symbols, and the integration of spirituality and psychology; the regions of the unconscious, including the location and role of the psyche and ego; and Jung's interest in trying to understand the physical-psychological relationship of synchronistic events in terms of modern physics.

Chapter 6: Teilhard's Sense of Collective Consciousness

Summary: This chapter summarizes the life work and experiences of the scientist-priest Pierre Teilhard de Chardin, the new sense that opened up his perception to the collective consciousness of what he termed the noosphere, and the role of this new sense in the evolution of human consciousness on the planet.

Key concepts and themes: Teilhard's inspiration leading to the development of what he termed a new science, hyperphysics, a physics of consciousness; his experiences and the opening of a "new sense" on the front as a soldier in World War I; how he discovered the the reality of noosphere; and his clear description of the arc of the future evolution of consciousness.

Part 3: Hyperphysics of the Noosphere

Overview: These chapters provide a detailed examination of the hyperphysics of the noosphere as developed by Teilhard as an extension of physical science.

Chapter 7: Hyperphysics: Speculative Physics beyond Material Science

Summary: This chapter reconciles physics with metaphysics, and "natural science" with "supernatural science." Through a more detailed examination of hyperphysics and the noosphere it provides a scientific basis for Teilhard's energy model of consciousness.

Key concepts and themes: Hyperphysics as an extension of physical science; the evolutionary arc of consciousness; the noosphere as an evolving manifestation of the global mind, a collective repository of individual experience.

Chapter 8: The Energetic Noosphere

Summary: This chapter offers a technically detailed map of energy as the matrix of consciousness the driver of evolution. Superposition of this living, communicating energy flux (holoflux) is shown to create and sustain various collective planetary *noospheres*.

Key concepts and themes: Human energy as central to human and noospheric consciousness; evidence of direct interaction of the electromagnetic energy of the geomagnetosphere with human consciousness; isospheres of consciousness

Chapter 9: The Superposition of Consciousness

Summary: This chapter discusses how individual centers of consciousness can resonate with other centers. It explores the mechanism of Teilhard's "new sense" in the evolution of the noosphere and provides a deeper dive into the layered experiences that make up our conscious awareness. This chapter details a topological map of the interconnectedness of the entire universal process in time and space, and discusses the superposition principle, which implies that the frequency domain aids in merging these separate but simultaneous experiences.

Key concepts and themes: Frequency resonance of the energy of consciousness; the mechanism of manifestation of the global mind and other collective minds; Fourier transform and inverse transform; the *superposition principle*; the role that the Fourier transform might play in brain/mind neurophysics; holography as frequency-superpositioned electromagnetic wave interference.

Part 4:
Sub-Quantum Consciousness

Overview: A more technical section than those previous, the material here offers the reader a deeper level of understanding of the cosmology and physiology of consciousness backed by the work of the brain scientist Carl Pribram and the quantum theorist David Bohm. Included is a discussion of how the universe (and consciousness) might operate in a digital mode (on and off) due to the discontinuity of time and space, and recent experiments in the physics of nondual communication.

Chapter 10: The Pribram-Bohm Holoflux Theory

Summary: The chapter is an examination of the map of consciousness developed over twenty years by Pribram and Bohm, and the mechanism of memory storage, and the dynamics of consciousness regarded as holoflux energy. It also looks at how the continuous mathematical operations of the Fourier transform links information (generated in space and time) with understanding (within the nonspatial, nontemporal dimensions of the implicate order), and the limits of space and time as granular, rather than continuous, in a holonomic metaverse of at least eleven dimensions of consciousness.

Key concepts and themes: Holoflux energy as dynamic consciousness linked in a web of multiple dimensions; the alignment of neurodynamics with quantum theory in the holographic operation of memory and mental processing.

Chapter 11: The Digital Universe

Summary: This chapter continues the description of the topology of the holoflux consciousness in the holonomic metaverse, using our knowledge of digital technology with an emphasis on scientific theories and measurements.

Key concepts and themes: Holograms, holopixels, and the universe as a multidimensional projection of holoflux consciousness; data storage and the cyclical nature of information feedback and circulation through the metaverse.

Part 5: Conclusions

Overview: This concluding chapter suggests how such knowledge might be used to establish the desperately needed connecting link between our previously isolated human psyches and the much larger collective psyches that may offer answers to our desperate search for guidance. Presented here are key concepts that provide the reader a more detailed understanding of consciousness and how an individual might establish links with wider networks beyond the human brain to channel information that addresses humanity's current challenges. Supported by contemporary research in clinical neuroscience, the topic of split-brain research and the bicameral mind is examined, followed by Gurdjieff's assertion that a human is a "three-brained being" with the ability to link awareness with a previously hidden, vast network of consciousness.

Chapter 12: Exploring the Hidden Dimensions of Mind

Summary: This chapter explores how the human mind operates as conscious holoflux, with information resonating in a multidimensional plasma throughout the human biocomputer (physical body and brain in spacetime).

Key concepts and themes: The structure of the psyche operating in the configuration of a human mind; split-brain research and differences between the cerebral hemispheres; consciousness awareness and practical applications.

KEY TERMS AND DEFINITIONS

Actual entity: In the philosophy of Alfred North Whitehead, an actual entity is an "occasion of experience," a fundamental unit of reality in the logically atomic sense that cannot be cut and separated into two other occasions of experience. This corresponds with Bohm's "holosphere," the smallest entity in spacetime, existing at the boundary between space and the other dimensions of the metaverse. Mystics would refer to this as an individual soul, *atman*, or *purusha*. It is the center and source of an individual conscious identity.

Explicate order: One of the several primary dimensional orders in the universe, as posited by the physicist David Bohm. The explicate order encompasses spacetime (the composite three dimensions of space and one dimension of time) and all manifestations within spacetime. The other dimensional orders Bohm identified as manifesting within the "implicate order," outside of the spacetime domain.

Holosphere: A spherical boundary that has the diameter of one Planck length (3×10^{-35} meter). This is the fundamental quantum unit that underlies space. The surface of a holosphere is an event-horizon boundary that marks the boundary between the dimensions of the explicate order (the four spacetime dimensions) and the implicate order (the hidden dimensions beyond spacetime).

Holoplenum: A continuous three-dimensional plenum of holospheres, the substrate out of which space emerges and through which spacetime-generated information flows into the implicate-order dimensions. This array of tiny holospheres have a similar function as do the pixels that power a video display screen, though the three-dimensional "holopixels" project the entire universe that we know in spacetime.

Holoflux: The multidimensional energy/consciousness/information flux that fills, spans, and links all dimensions (at least eleven and possibly many more); in the spacetime region, holoflux energy includes all energies: gravitational energy, electromagnetic energy, and the strong and weak nuclear forces.

Implicate order: The implicate order is Bohm's term for the nonlocal, nontemporal domain consisting of all dimensions that lie beyond those of space and time (quantum string theory's eleven dimensions and perhaps more). Bohm felt that the implicate order is "'deeper' and more fundamental than the explicate order," and is a primary location of consciousness.

Metaverse: The entire "whole of everything" that includes the universe of space and time, and of all that is beyond this universe; an integrated network of multiple dimensions of the single underlying consciousness that underlies and projects the whole of reality. This includes, in addition to space and time, at a minimum, an additional seven primary dimensions (as indicated by M-theory) as well as innumerable, perhaps infinite, virtual reality dimensions that are open to creation by consciousness and available for exploration, colonization, and codevelopment by projections of the omni-interconnected conscious metaverse.

Noosphere: An evolving collective consciousness into which innumerable actual entities (souls or instances of individual consciousness) are networked, linked into a higher configuration of consciousness without losing their sense of individual identity. It can be thought of as a collective group mind that transcends time and space. According to Teilhard de Chardin there may be innumerable noospheres throughout the universe and within individual planets, stars, galaxies, and clusters of galaxies.

Planck length: The smallest possible length in space as determined by the father of quantum physics, Max Planck, from the gravitational constant and the speed of light. The value of the Planck length is 3×10^{-35} meter.

Planck time: The smallest possible unit of time, determined by Max Planck from the gravitational constant and the speed of light. The value of Planck time is 5×10^{-44} sec.

Psychonaut: From the Greek ψυχή, *psychē* ("soul," "spirit," or "mind") and ναύτης, *naútēs* ("sailor" or "navigator"), thus a "sailor of the soul" engaged in exploration of oceans of consciousness that lie in normally "hidden" dimensions, accessed traditionally through the means of prayer, meditation, drugs, and psychophysical exercises.

Part 1
•
THE HIDDEN DIMENSIONS OF CONSCIOUSNESS

1
What Are the Hidden Dimensions?

In a 2022 paper on astrophysics, Bernard Carr, a professor of mathematics and astronomy at the University of London who had studied directly under Stephen Hawking at Cambridge, wrote of the possibility of hidden dimensions beyond spacetime:

> The Universe may have more than the three dimensions of space that we actually observe, with the extra dimensions being compactified on the Planck scale (the distance of 10^{-35} meters, at which quantum gravity effects become important), so that we do not notice them. In particular, physics has revealed a unity about the Universe which makes it clear that everything is connected in a way which would have seemed inconceivable a few decades ago. The discovery of dark dimensions through particle physics shakes our view of the nature of reality just as profoundly as the discovery of dark energy through cosmology.[1]

It is generally assumed that our five major human senses (vision, hearing, smell, taste, touch) cannot perceive these extra dimensions referred to by Carr, and it appears that until recently these additional dimensions of reality have remained hidden from the scientific community. But the unexpected discovery of numerous additional dimensions has been the direct result of careful mathematical analysis of recent high-energy particle-physics experiments conducted in extremely expensive devices that are commonly called "atom smashers." One of these early particle collider-accelerators, called the "bevatron," first becoming operational at the

Fig. 1.1. The Bevatron collider, 1963; Berkeley Lab.

University of California at Berkeley in 1954 (fig. 1.1), opened the doorway to modern dimensional research.

The largest and most powerful of all colliders at the present time is the Large Hadron Collider (fig. 1.2) at the European Organization for Nuclear Research (Conseil Européen pour la Recherche Nucléaire, or CERN), within which the particle streams are accelerated for a distance of seventeen miles within a vacuum tunnel buried six hundred feet deep below the soil of the Switzerland-France border near Geneva. In a typical experiment, two streams of protons, streaming in opposite directions within the central tube, are accelerated to the maximum velocity possible using a series of two thousand superconducting electromagnets, some weighing up to twenty-eight tons.

During a single experiment, 200 MW of electricity is drawn from the French electrical grid, about one third of the entire power used by the nearest city, Geneva. This massive power forces the accelerating particles in a narrow stream through almost 1600 focusing magnets, some as long

Fig. 1.2. Magnets within the Large Hadron Collider, near Geneva.

as fifteen meters in length, until the two streams reach speeds approaching the speed of light. When maximum velocity is detected within the collider, there arrives the moment when the magnet controls abruptly shift the streams into a direct collision path with one another.

At the moment of collision, high-speed electronic imaging devices (fig. 1.3), each one fifteen meters in diameter and weighing close to 15,000

Fig. 1.3. A muon solenoid detector at CERN.

tons, take high-resolution snapshots of the results, to reveal tracks coming out from the disintegrating protons. During most experiments about fifty proton pairs collide, shattering into smaller quantum components at the moment of collision. Mathematical analysis of the trajectories of these sub-quantum spinoff particles supports the suspicion that there must exist *at least* seven dimensions *beyond* the three dimensions of space and single dimension of time.

What Are the Hidden Dimensions?

Understandably, few people, other than mathematicians, physicists, and psychonauts, spend much time thinking about dimensions. This is quite understandable. The concept of a *dimension* is clearly abstract and infrequently encountered in everyday conversation. Likely first encountering the term in basic classes of geometry in primary school, most people are familiar with the three dimensions of space, the x, y, z axis lines depicted in graphs (fig. 1.4).

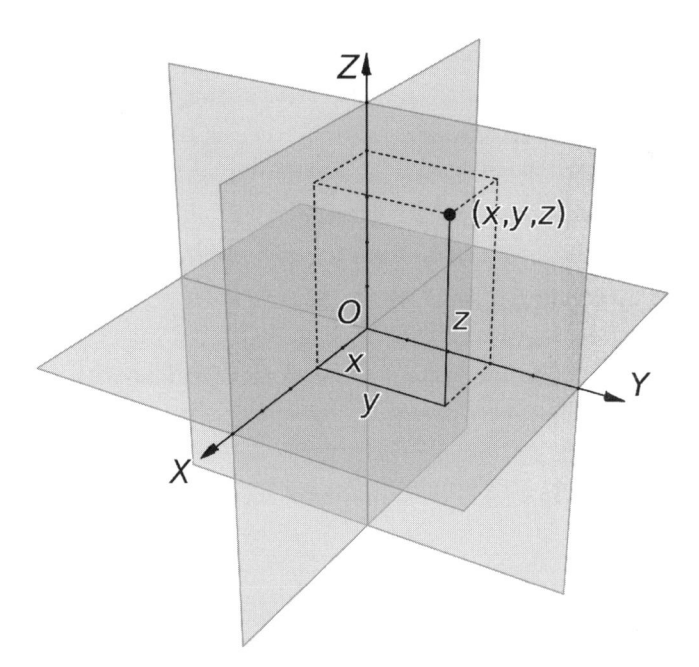

Fig. 1.4. The three dimensions of space (x, y, z).

The dimension we call *time* is understood to be a single dimension, while the other three familiar dimensions exist in what we term "space." Everyone accepts the fact that we all live in these four dimensions, and only science fiction writers and a very few scientists have previously considered the possibility that there might really exist additional dimensions about which we know almost nothing.

Mathematicians refer to this mesh of four dimensions, three of space and the one of time, as "Euclidean space." Physicists use a shorthand term, frequently referring to the four dimensions simply as "spacetime."

Now, however, with the advent of *string theory,* developed through mathematical analysis of observations in particle physics, there is growing acceptance that there exist additional dimensions beyond traditional Euclidean space. This has become perhaps one of the strangest and least known discoveries in the field of quantum physics. Since the first paper on string theory was published in the 1970s, string theory has become one of the most widely discussed approaches among physicists in their efforts to interpret results from high-energy particle experiments.

String theory holds that the universe in its entirety must have ten distinct dimensions. A slightly newer version called "M-theory" (in which the M is said to stand for "mystery"), includes gravity as a separate dimension, thus considering the possibility that there are eleven distinct dimensions. In fact there is even a newer, rather esoteric version of string theory called "Closed Unoriented Bosonic String Theory" that proposes twenty-six dimensions! For consistency in this book, we will assume that we are discussing the exploration of eleven distinct dimensions of M-theory, seven hidden dimensions beyond the four of our familiar spacetime.

The strings discussed in string theory (sometimes called superstring theory) are the smallest identifiable objects to manifest within our cosmological three-dimensions of space. In some sense they can be compared to strings on a musical instrument, however, each string in string theory is so miniscule that it would appear to us as a point-like particle down at the limit of space: the Planck length of 10^{-35} meters, pure energy, much smaller than an atom. They have been called "micro black holes" by Stephen Hawking, and Planck-length holospheres by David Bohm. The importance of strings lies in the fact that they are thought to be the building blocks that are responsible for bringing into existence and sustaining all of the different particles that are currently known to be produced in high-energy particle-collision experiments (e.g., electrons, neutrinos, quarks, bosons, etc.). According to string theory, all identifiable quantum objects depend directly upon (and are produced by) these "strings" that are vibrating at different frequencies in various dimensions that are thought to exist beyond the dimensions of space and time.

According to this theory, each of these strings can vibrate in a different mode, in unique sets of a multitude of frequencies, much in the same way one can produce different notes from a string on a guitar or cello. Our familiar dimensions of time and space do not give the strings enough room to vibrate in all the ways that are required in order to fully express themselves as all the varieties of particles known to the world of quantum mechanics. Space and time are just too constrained! One physicist has written, "the strings don't just wiggle, they wiggle hyperdimensionally."[2] The number of unique vibrational configurations of a single quantum string vibrating in ten dimensions has been calculated to be around $10^{2,000,000}$, which, assuming such vibrations may be modulated with information of some sort, offers an inconceivable richness of possibilities for communication within the metaverse with such strings vibrating within these multiple dimensions simultaneously!

String theory has provided the mathematical precision to successfully predict outcomes of high-energy particle-collision experiments, while also pointing to the reality of the additional and previously unsuspected dimensions. In attempting to comprehend these newly discovered dimensions, it has been speculated that these hidden dimensions must exist either "outside of" the expanding universe (the one that can be viewed and measured), or that the hidden dimensions might lie "above, within, or below" our spatial cosmos, that is, outside of space in regions vastly smaller than even a proton, below the Planck length limits of space that are known to be below 10^{-35} meters, most likely in what the physicist David Bohm termed "the implicate order."

Unfortunately, the use of the word *string* can be misleading as it is strictly a mathematical term, one that should not be confused with a string made of twine used to wrap a package. A "string" of one dimension would not be visible on a two-dimensional plane. Instead, these one-dimension "strings," frequently used in calculations by quantum physicists, might be better visualized as extremely small points or pixels, existing only at the bottom of the scalar universe of space, at the boundary between space and the hidden dimensions that lie beyond the three-dimensional region we refer to as space. These strings might be better visualized as micro black holes rather than as an infinitely small piece of twine or a fiber of lint. As discussed later in this book, the strings discussed by quantum physicists may be actually equivalent to the following concepts:

- the micro black holes or "holospheres" that form a plenum in David Bohm's model of the universe.
- the "Omega point" in Teilhard de Chardin's geometry of the universe;
- the "actual entities" of consciousness proposed by the mathematician-philosopher Alfred North Whitehead.*

At present, these string entities remain undetected by human laboratory equipment, as they are many, many magnitudes of size *smaller* than atoms or electrons (the diameter of a proton is known to be 10^{-14} meters while a string is predicted to be 10^{-35} meters.) Though incredibly small, these strings are thought to establish the structure of the universe. David Bohm considered them to exist at the bottom of space as a *plenum*, much as philosophers from Aristotle to Descartes accepted this concept of the plenum, believing that there are no empty spaces, but that the entire foundation of the universe is built upon a plenum of tiny close-packed spheres (or strings in the language of quantum physicists).

Experimental Proof of Seven Hidden Dimensions

The wide acceptance of string theory stems from the fact that it is the only mathematical model that can explain innumerable unexpected results from data gathered over years of experiments in colliding fundamental particles such as protons and electrons in particle accelerators (cyclotrons, synchrotrons, bevatrons, tevatrons, etc.). As noted above, the largest of these types

---

*For Whitehead: "actual entities" are the fundamental building blocks of reality, the "ultimate facts that constitute the world." Nothing exists beneath or beyond these actual entities; instead, they serve as the foundation for all of reality. Consciousness is a product of the ongoing process of becoming that defines the existence of all actual entities.

Fig. 1.5. Inside a collider tunnel.

of devices is the Large Hadron Collider near Geneva, Switzerland. The largest machine ever built by human beings, the LHC uses hundreds of enormous superconducting electromagnets—each weighing seven tons and cooled with liquid helium—that accelerate charged subatomic particles, primarily protons, to the astounding speed of 99.999956 percent of the speed of light. The speed at which such a particle moves is directly proportional to its raw energy at the moment of collision.

Modern colliders are designed as rings that have circumferences that allow the beams of accelerating particles to gain enormous speed and energy, reaching slightly more than 99.99 percent of the speed of light, before they are made to collide with particles moving in the opposite direction. Details of the Large Hadron Collider can be seen in the central area of figure 1.5.

The current LHC took ten years to construct at a cost of almost five billion dollars, a cost shared by several countries. It was first fully energized in 2008 while the gathered crowd of physicists listened to the amplified sound of Beethoven's "Ode to Joy." High-energy particle collisions are measured in energy units of the TeV (trillion electronvolts). One TeV of energy accelerates particles (usually protons, which have diameters of 10^{-15} meters). Two streams of protons are accelerated in opposite directions in close proximity to one another, going faster and faster with each acceleration through the seventeen-mile circular ring.

28 • The Hidden Dimensions of Consciousness

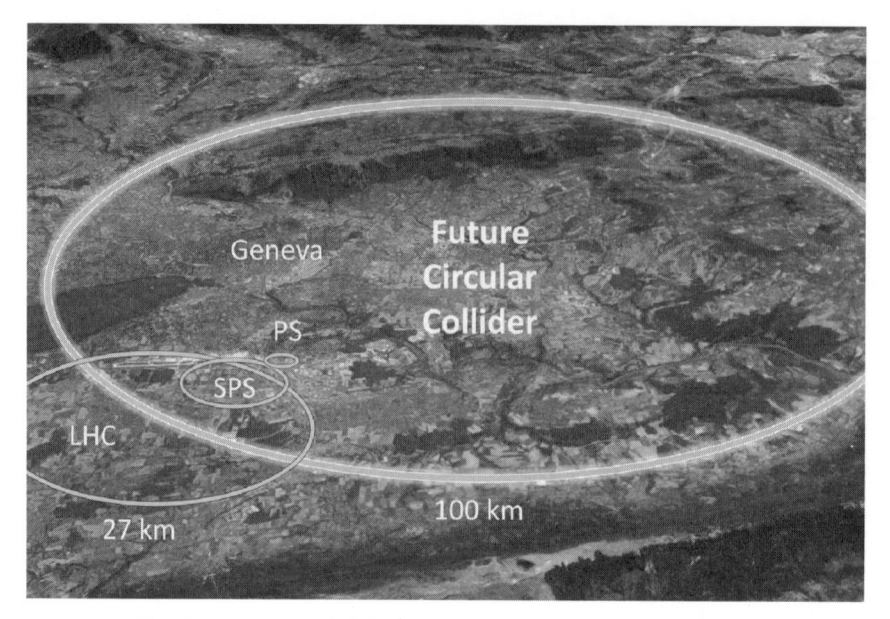

Fig. 1.6. Current LHC and the Future Circular Collider.

The LHC has a circumference of slightly more than 17 miles (rounded up from 16.7) (27 km), yet a new circular collider (fig. 1.6) is currently being built next to it that will have a total circumference of 64 miles (100 km).

During a typical experiment within the collider, the two streams of protons whirl around the circumference until they reach peak acceleration. At that moment the electromagnetic guiding fields are minutely adjusted so that the two streams collide, smashing head-on into one another. The resulting impact generates an enormous amount of energy, the most recent record being 13.6 trillion electronvolts (13.6 TeV). An enhanced collider is expected to generate collisions of over 100 TeV.

In early experiments, much to the surprise of physicists, ten subdimensional particle tracks were observed coming out of the collisions at 90 degrees to the tracks of the original particles, indicating the existence of underlying additional dimensions never previously suspected. Over the past few decades, particle accelerators have been improved, always with the goal being to generate higher and higher collision energies.

In analyzing the mathematics that would explain (and predict) these ten unanticipated particle tracks, it was discovered that there must indeed *be an ontological reality* to the additional dimensions, a reality clearly beyond the four dimensions of space and time. It is clear that all of these dimensions are somehow interconnected and thus they must all intersect at a single point in hyperspace, perhaps Teilhard de Chardin's "point Omega." As the astrophysicist Paul Sutter describes it:

> There are a lot of different ways to wrap those extra dimensions in on themselves. And each possible configuration will affect the ways the strings inside them vibrate. Since the ways that strings vibrate determine how they behave up here in the macroscopic world, each choice of manifold leads to a distinct universe with its own set of physics.[3]

In more ordinary terms, a "one-dimensional string" can also be described as a "point." But even the word *point* is open to at least two interpretations:

- In Euclidean geometry a point is a location with no size, that is, no width, no length, and no depth. It only has position.
- Modern mathematicians define a point as a zero-dimensional mathematical object that is specified in *n*-dimensional space.

Mathematical physics defines these strings to be energy processes that occur at the very lowest limit of the three spatial dimensions, down at the scalar bottom limit of 10^{-35} meters, below which space itself has no meaning and no longer exists. String theory predicts that all of the other seven dimensions are hidden from spacetime by being somehow "rolled up" within a miniscule boundary (the event horizon) of tiny regions that might best be visualized as spheres with diameters equal to the Planck length (roughly equal to 1.6×10^{-35} m or about 10^{-20} times the size of a proton). This spherical region at the bottom of space is called a Planck holosphere, also known as a micro black hole, or simply a holosphere, and will be discussed in later sections of this book.

The renowned quantum physicist Stephen Hawking believed that string theory might very well grow into a "theory of everything," an integrated model for describing the makeup of the universe. Yet even Hawking thought that his own views on the subject had become somewhat "spooky," so much so that in 2012 he warned against firing up the

30 • The Hidden Dimensions of Consciousness

Large Hadron Collider to manipulate strings in an experiment seeking to probe the hidden dimension from which emerges the Higgs boson.

Known as the "God particle," the Higgs boson was predicted to emerge from the Higgs field during an extremely high-energy collision of protons (1.6×10^{-15} in diameter). The experiment Hawking was worried about was the attempt to prove the existence of a hidden dimension predicted by string theory that can only exist outside of time and space, yet paradoxically permeates all of space. From the hidden dimension of the Higgs field, the Higgs boson was theorized to emerge, bringing dark

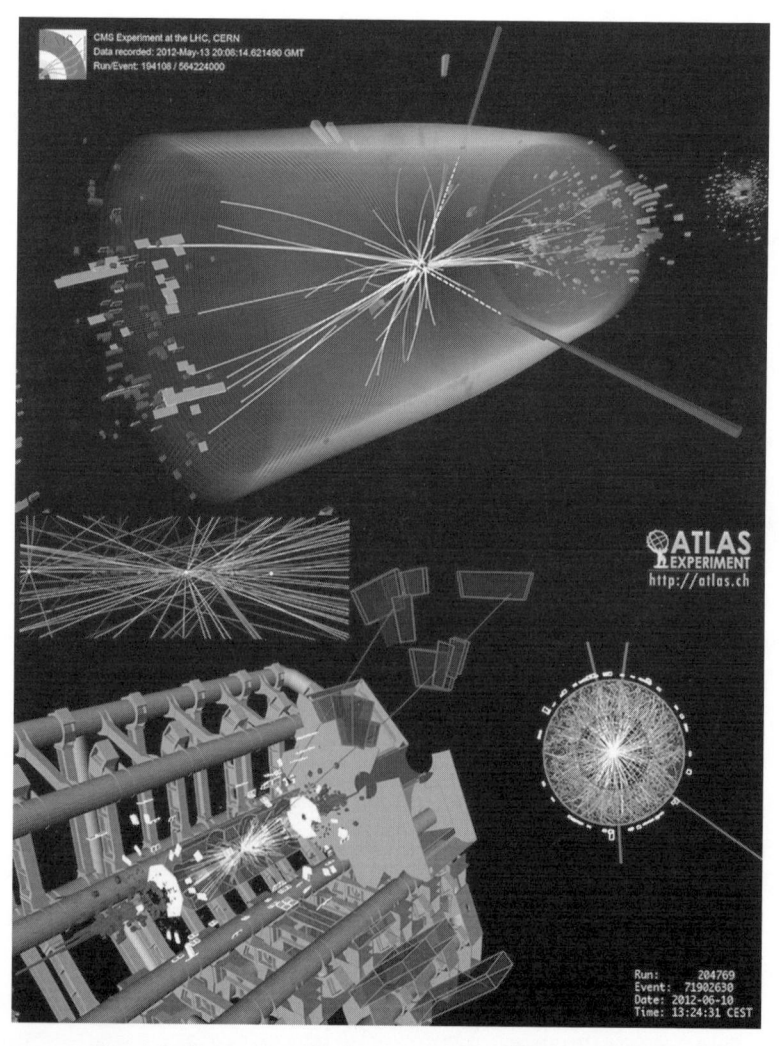

Fig. 1.7. Higgs boson collision image within the LHC.

energy mass into our spacetime universe. The observation of a Higgs boson emerging from this hidden dimension would confirm the existence of the Higgs field, and thus prove string theory correct. Yet Hawking was so alarmed that he published the following warning shortly before the scheduled Large Hadron Collider experiment would attempt to validate string theory by smashing proton streams with enough energy to cause an actual Higgs boson to emerge. Hawking stated:

> The Higgs potential has the worrisome feature that it might become metastable at energies above 100bn giga-electronvolts (GeV). This could mean that the universe could undergo catastrophic vacuum decay, with a bubble of the true vacuum expanding at the speed of light. This could happen at any time and we wouldn't see it coming.[2]

Yet his fears were unfounded. On July 4, 2012, using the Large Hadron Collider (LHC), three hundred feet below ground on the French-Swiss border, two streams of protons, speeding in opposite directions around a five-mile diameter collimated pipe ring, were made to collide head-on from opposite directions at extremely high energies. To the delight of the hundreds of physicists monitoring the test around the world, the collisions quickly produced evidence of the emergence of the long sought-after Higgs boson from the hidden dimension of the Higgs field. The "God particle" had finally been detected! Space and time are not alone among the dimensions! And to Hawking's relief, the universe had not collapsed! Figure 1.7 shows several artistic views of the emergence of a Higgs boson from the hidden dimension within the Large Hadron Collider.

The bright linear trails visible in the image are the paths of particles emerging from the direct head-on collision of the two streams of protons that are speeding in opposite directions at enormous energies.

If the universe, or the wider metaverse itself, does exist emerging from hidden dimensions, what then might be the nature of this metaverse? The following chapter discusses the universe, the metaverse, and their relationship to the psyche and human mind.

2
Universe or Metaverse?

The word *universe* is itself an outmoded term. "Universe" has been historically defined as the entirety of space and time and their contents, including planets, stars, galaxies, and all other forms of matter—that which can be observed using our four physical senses. However, that has now been superseded by a new word, the *metaverse*, a term that implies a much larger reality than that perceived by our ancestors. Recent developments in physics, mathematics, and neurobiology point to actual dimensions that lie far beyond our current sensory systems. Surprisingly, these findings have been found to support the contention of mystics and a growing number of psychonauts that there exist hidden dimensions, vast new regions of reality open to human exploration. Traditional contemplatives insist that it is within such esoteric regions of awareness that significant knowledge and guidance can be found. Within the written and unwritten traditions of many cultures, these experienced contemplatives assure us that, with training and practice, a solitary individual can come into contact with and perceive and experience multiple domains in dimensions beyond those of sight and sound. More importantly, we are told that it then becomes possible to "converse" with the nonhuman intelligences that populate these otherworldly, parallel dimensions.

Archetypal Forms and the Metaverse

The scientist and Tibetan scholar B. Alan Wallace describes the role of archetypal forms in the process of deep meditative concentration:

In accomplishing meditative quiescence, the mind shifts from the physical realm in which experience is dominated by the senses to a realm of *archetypal* forms, called the "form realm." The form realm is beyond human senses, culture, and conditioning, but is accessible while having a human body. This, I expect, is the realm in which the laws, patterns, and mathematical truths of nature exist. The Buddhist claim is that a human being can, by means of deep meditative concentration, access the realm of archetypal forms, and even formless dimensions of boundless space, consciousness, nothingness, and a state of being that transcends the concepts of discernment and non-discernment.[1]

When Wallace describes the ability to "access the realm of *archetypal* forms," he is referring to *archetypes*, a major subject in several of Jung's essays, including "Archetypes and the Collective Unconscious," that also supports Teilhard de Chardin's concept of the collective *noosphere*. The word *archetype* is a term first used by Irenaeus, a scholar and priest who lived in the early second century who was widely known as an authority on Gnosticism, a collection of religious ideas and practical techniques for accessing spiritual knowledge in the higher worlds. Irenaeus taught that there exists an invisible world beyond space and time that projects the forms and patterns of visible things on Earth: "The creator of this world did not fashion these things directly from himself but copied them from archetypes outside himself."[2]

Irenaeus, Jung, and Wallace all have in common the use of the term *archetypal form*. Jung understood the archetypes to be dynamic, living patterns of consciousness that he referred to as *psychoids*. Archetypal psychoids have the potential to transform human consciousness through an influx of new and empowering content. Through religious, magic, and commercial images and rituals, the archetypal form can establish links between a human psyche and the archetype. Thus highly charged images (Cross, Star of David, Yantra, American Flag) can act much like an address, telephone number, or internet hyperlink—allowing the archetype to connect immediately to the psyche in spacetime.

Once a link is established between the psychoidal entity in its hidden dimension and the human psyche (whether individual or collective), a process begins that alters the spacetime human psyche. Similar to the downloading of a new operating system or software application to a personal computer, the psychoids can download resonant templates that may radically influence a wide variety of psyches: individual, family, tribal, political, national, ethnic, or the collective psyche of a particular religion, political faction, or belief system. The archetype downloads a template or structure with which the psyche can begin to identify, and thus the influence of the archetype can grow.

This idea can be seen corroborated by the molecular biologist Rupert Sheldrake, who calls the phenomenon *morphic resonance*.*

Sheldrake posits that memory is inherent in nature and that natural systems inherit a collective memory from all previous instances of their kind. He also proposes that what he terms morphic resonance is a mechanism for the widespread observation of telepathy-like interconnections between living organisms. While an archetypal template of an unfolding cherry blossom can be viewed as a projection from the archetype of a cherry blossom into space and time, more complex archetypal patterns unfold more complex spacetime processes in the human psyche, such as patterns of experiences such as love, altruism, courage, ingenuity, and morality.

The physics of the late David Bohm supports this idea that our world in space and time is a projection of archetypal forms from transcendent dimensions:

> Bohm proposed that the understanding of motion be viewed instead as a cyclic process of projection, injection, and re-projection. The archetypal form is projected *from* a state of potentiality into matter that then is imbued with experiential knowledge of the material world, and then subsequently enfolded back *into* the domain of forms, and the re-projected. Bohm's implicate order can therefore be understood as a model of an evolutionary metaphysics.[3]

---

*Sheldrake's theory of *morphic resonance* supports the idea that there exists a collective memory not stored in material traces within the brain, but exists separately elsewhere as an energy force field, acting as an information template to guide the development and behavior of an organism as it grows and evolves. The template (archetype) itself is affected by developments in the organism as it evolves in space and time and through resonance transfers new information back into the archetypal source template.

Psychoids, Archetypes, and Instincts

Support for the ideas of Wallace's archetypal forms and Sheldrake's morphic resonance can be found in Jung's observation that there are vast regions of consciousness that act to directly influence our healthy (or unhealthy) physical growth and mental development. Numerous alien (nonhuman) psychoids affect us though they exist and operate outside of the normal waking awareness of our "ego." Jung calls these regions "unconscious" with respect to our ego; we are normally unaware of these psychoids, only because we are unconscious of the many processes affecting us that are going on continually within the universe. Psychoids, both instinctual and archetypal, are in themselves forms of consciousness (conscious entities), though clearly not human consciousness as we know it.

Jung sees our individual psyches as being directly influenced by a cosmos that is filled with psychoids of innumerable kinds residing in multiple dimensions even beyond space and time (Jung uses the term *transcendent* for these entities). Many of these centers of consciousness, existing in various channels or bandwidths of consciousness, guide our physical and psychological development and act as templates. They commune with us through direct resonance with them (a phenomenon of spectral frequency entrainment). However, there are also dangerous psychoids that can influence the human psyche in dark and destructive ways. In a sense, the traditional religious view that there are "angels and demons" is perhaps quite fitting as an acknowledgment of an experienced reality.

In his later works, Jung develops the analogy of the wide range of consciousness as a frequency spectrum (fig. 2.1). In the higher frequencies (ultraviolet and blue, in the case of the visible light spectrum) can be found the archetypes, while the lower-frequency region of the spectrum (red and infrared) is the domain of instinct psychoids.

In communications technologies it is well known that higher frequencies offer the capacity to handle more data and at a greater speed than lower frequencies. In Jung's map of consciousness, the lower-frequency bands can handle relatively simple "instinctual" subprograms, while by contrast the higher-frequency bands support the more complex activities of archetypal psychoid information processing.

It is clear to both Jung and Wallace that the wider dimensions of higher consciousness, containing the energy patterns of consciousness

Psychoids:
Archetypes   # Instincts

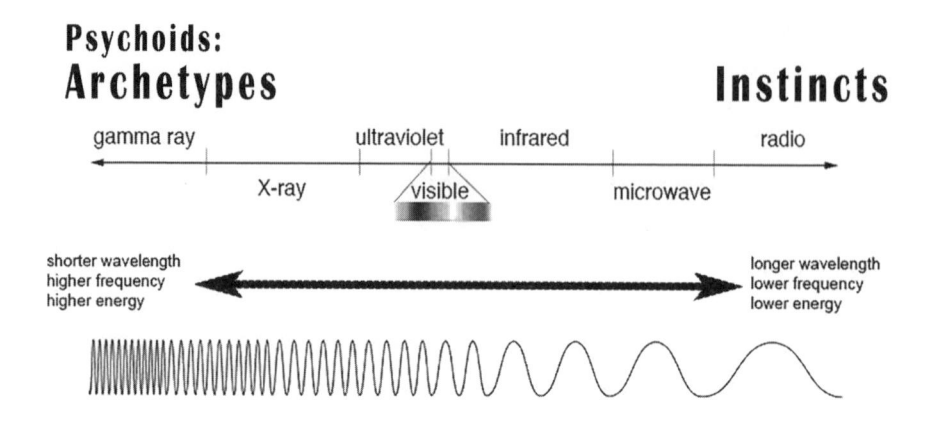

```
gamma ray          ultraviolet    infrared            radio
         X-ray            visible         microwave
```

shorter wavelength   longer wavelength
higher frequency   lower frequency
higher energy   lower energy

Fig. 2.1. Psychoids on the frequency spectrum.

that they both refer to as *archetypes*, are not usually perceived by an "I-consciousness," and are generally inaccessible to a typical human's daily waking consciousness. Jung uses two different terms to identify influences from "lower" and "higher" bandwidth regions of consciousness. For entities in the lower instinctual range Jung uses the term *psychoids*, while for the higher collective consciousness he uses the term *archetypes*. In traditional cultures, the use of symbols, mythology, art, and ritual were developed over the centuries and millennia to establish, maintain, and strengthen links to these archetypes, viewed as the many "gods" and "angels" by cultures throughout the globe. With the advent of material science, the public focus of consciousness moved away from traditional religious activities that previously were able to link the individual "I-consciousnesses" into a hierarchy of various collectives (e.g., of family, tribe, region, nation, etc.) through the maintenance of a living resonance with organizing archetypal energy patterns. Sheldrake would see this as a case of morphic resonance, though his work is focused more on biology than psychology.

The traditional links (for example, links filled by the role of the Catholic Church in the Middle Ages) to powerful archetypes through symbol and ritual resulted in a gift of harmonizing patterns that gave the individual "I-consciousness" a sense of belonging that at the same time, powered by the archetypes, fostered other-dimensional qualities such as hope, love, altruism, modesty, humility, and repentance for error. The church maintained the active templates of powerful mythological symbols and revered stories that in turn worked to guide human egos to live in harmony with the eternal process of change unfolding in space and time. Today these links have been severed and replaced with rampant materialism and a focus only on space and time, ignoring the hidden dimensions of consciousness.

It is clear that traditional links to the archetypes through monocultural religious symbolism, liturgy, art, and mythology, have been lost. We have lost touch with the many "gods" of the Greek and Roman traditions, for example, which are the archetypes that populated what Jung called "the collective unconscious." Without such links, our collective psyche has nothing else to focus upon except the material, and it is paying the price. The modern collectives are fragmented into competing groups and cultures—more specifically focused upon wealth and sensory delights, digital devices, and physical entertainment—rather than working to maintain and develop healthy interactions with archetypal patterns. No longer widely accessible are the knowledge patterns (existing everywhere beyond space and time) that in the past brought harmony and healing to the collective psyche.

In his work as a psychoanalyst, Jung strove to help individuals identify and then reconnect their "I-consciousness" with various missing aspects of consciousness that may have been exerting pathological influences on the patient's "I-consciousness."

Similarly, the original intent of all religious belief systems was to reconnect (*re-ligiō*, "to bind; to tie back again") with the higher states of normally hidden consciousness, to establish links with the missing invisible dimensions of our souls. To "reconnect," individuals must learn techniques with which to explore transcendent dimensions in order to commune with the realities of living conscious entities, or *psychoids*. The term *metaverse* (*meta* as in transcending the universe) is a more accurate and useful term than *universe*, which is conventionally taken to be everything in the four dimensions of space and time. The term metaverse, however, includes the entire reality of conscious dimensions at every level, beyond and within space and time.

Current technical support for the existence of these nontemporal, nonspatial dimensions can be found in modern superstring theory, a widely accepted model in high-energy particle physics. An additional map can be found in the work of the physicist David Bohm, who pioneered the idea of the *implicate order*, a term Bohm used to signify the existence of the hidden dimensions, as contrasted to the *explicate order* of our familiar space and time.* As discussed in the previous chapter, superstring mathematics of M-theory predicts eleven dimensions. Einstein's general theory of relatively treats time as a single dimension while considering space to be made up of three additional dimensions: height, width, depth (or x, y, z). Einstein's equations treat them, taken together, as four-dimensional "spacetime." We now know that there are, in fact, other dimensions beyond the four mapped by Einstein. The proponents of M-theory point to eleven dimensions that are now considered to be "hidden" from our spacetime senses. Many quantum physicists and mathematical cosmologists have published similar findings indicating that there must be *at least* seven dimensions beyond space and time, while a new theory called Closed Unoriented Bosonic String Theory interprets results from recent test as indicating the existence of 26 dimensions (and an even smaller group speculates there may be an infinite number of dimensions beyond space and time). Clearly, our scientific (and psychic) explorations can no longer be limited to four dimensions!

String Theory and Knowledge of the Higher Worlds

With the recent discovery of hidden dimensions, human exploration has expanded into a broader metaverse of consciousness. The virtual realities created within software networks might only be the training ground for future expeditions into these unknown realms. Just as early fish, equipped with rudimentary fins, ventured onto the shores of a new dimension called land, we are now beginning to explore these new dimensions through contemplative introspection. However, as the fish eventually evolved legs and wings to explore and map the lands beyond the shore, we too should be able to develop new methods of navigating these dimensions. With the help of traditional methods, like yoga and psychedelic drugs, we must begin to develop programs to find, explore, and map these new worlds.

---

*See "Quantum Theory as an Indication of a Multidimensional Implicate Order" in Bohm's book, *Wholeness and the Implicate Order* (1980).

Modern technologies such as AI can be used to help us create maps of these unknown dimensions of consciousness. With such guidance we may be able to mine the hidden dimensions described as "the higher worlds" by our ancestors, to obtain knowledge and understanding to help humanity, individually and collectively, move safely into a richer integration of consciousness in harmony with the world around us and within us.

Mystics have often spoken of "higher worlds." In all cultures the exploration of these worlds has frequently been described in the accounts of individual shamans, yogis, mystics, and psychonauts. The renowned Austrian scientist and mystic Rudolf Steiner, for example, described the situation in his book, *Knowledge of the Higher Worlds and Its Attainment*:

> There slumber in every human being faculties by means of which he can acquire for himself a knowledge of higher worlds. Mystics, Gnostics, Theosophists — all speak of a world of soul and spirit which for them is just as real as the world we see with our physical eyes and touch with our physical hands.[4]

What then are these "faculties" by which, Steiner tells us, we can awaken to the perception of higher worlds? They are definitely not ordinary vision or hearing or taste or touch. We do not even have any commonly accepted names for such new sensory systems other than perhaps "extrasensory perception," while religions and mystical traditions refer to such an experience as "a sixth sense," "having a vision," "acquiring a new sense," "having a revelation," and so on.

What is it that might be found in these hidden dimensions that could possibly offer us answers to our basic questions or perhaps offer advice and even possible assistance in our lives, individually and collectively? The answer is that these hidden dimensions are populated with their own unique forms of consciousness, as predicted by panpsychists, who believe that consciousness is the basis of all existence and is the bedrock of the cosmos. If there are other dimensions, and it is becoming more apparent that there are, it is reasonable to expect there are conscious entities inhabiting those dimensions and with which we might be able to enter into communication during explorations of those hidden dimensions.

Folklore as well as centuries of written accounts support both the existence of such inter-dimensional beings and the ability of humans to converse with them. Those who explored hidden dimensions had to coin new names to describe the various experiences and entities they encountered within the hidden dimensions. From these courageous psychophysical pioneers of inner space (many accused of witchcraft) we have inherited a plethora of descriptive words: gods, angels, archangels, demons, ghosts, fairies, specters, seraphim, cherubim, leprechauns, *rakshasas*, thrones, dominions, doppelgängers, shapeshifters, and so on. Rudolf Steiner frequently lectured and wrote of a spirit he called "the Archangel Michael" with which he frequently communicated, telling us that this "spirit" was one of the major conscious entities caring for planet Earth.[5]

What then might be the new sensory systems that we need in order to enter and explore the hidden dimensions, the higher worlds? Perhaps the faculties that Steiner writes about are the vestiges of ancient sensory systems that have atrophied over millennia through lack of necessity, or alternatively, might they not be completely brand-new "growth-tips" of the evolving human species? The scientific community has not until now found the interest to investigate the dimensions beyond time and space owing to cultural bias, not lack of resources. Yet shamans, yogis, monks, and other mystics in all cultures on the planet have explored other dimensions over the millennia, during which each has developed their own sophisticated methodologies to enhance their human physiological sensory systems in order to activate the capacities that allow them to explore the transcendental metaverse through immediate experience.

Only recently have a growing number of scientists, at significant risk to their reputations, begun to focus their efforts on exploring new dimensions of consciousness, often assisted by ingestion of entheogens and/or various traditional psychophysical exercises (e.g., yoga, tai chi, breathwork, mantra, esoteric visualizations, etc.). It may be possible that our *Homo sapiens* physiology, shaped by 300,000 years of evolution, may ultimately prove to be a more effective tool for directly exploring the metaverse than any hardware. It is also conceivable that electronic devices, perhaps assisted by AI, could eventually be harnessed to guide consciousness into entering and navigating these esoteric dimensions, much as molecules of peyote, LSD, and other psychotropic drugs have been found effective in exploring the hidden realms of awareness.

Physical science has brought us many amazing devices, but it may be the union of science and mysticism, physics and metaphysics that now hold the key to a better understanding of who we are and how we might be able to solve the many problems we now face in our lives and on our planet.

Networks of Consciousness

According to Donald Hoffman, neural networks exist exclusively within the bounds of spacetime, while conscious agents make up networks of consciousness that operate beyond spacetime. According to Hoffman's mathematical model, conscious agent networks operate *outside* of our four-dimensional cosmos, using spacetime primarily as a medium for creation and an interface for communication. Shamans, mystics, and psychonauts describe the existence of vast oceans of consciousness, networks of living webs of awareness that exist in dimensions far beyond those revealed by our normal human senses. Those who have developed this new ability assure others that with sufficient effort and practice, every human can waken to these innate sensory abilities, and that with sufficient will and effort, anyone can discover how to open one's "inner eye" to explore the mysteries of the metaverse.

The most classic approach to linking to transcendent networks of consciousness can be seen in the myriad varieties of meditation, contemplation, and prayer practiced within religious traditions worldwide. Prayer, both audible and inaudible, has been found pragmatically to be a powerful way to access invisible dimensions and to open channels of communication with psychoids that exist both within and beyond space and time.

Now, let us take a broader cosmological perspective. The essence of consciousness is that it is able to communicate; to exchange information. So how does the radiant cosmos communicate in our universe, and perhaps in the wider metaverse? The modern materialist paradigm holds that consciousness is a recent phenomenon, emerging into the universe by some kind of accidental process from neurons, brains, and microtubules embedded in meat. But what actually arose first in the cosmos? Cosmologists know what came first: not the brain, not the microtubules—but glowing, radiant electromagnetic energy. Galaxies, stars, and planets are structurally similar—there are amazing recurrent patterns in the universe. Accordingly, there must be some some *thing*, some repository of consciousness laden with information that must be transmitting the structural patterns that recurringly develop into stars and galaxies and planets in spacetime.

So just how does this radiant energy communicate? How does the Sun communicate within itself or with its planets or with a myriad of conscious entities that may exist within the metaverse? How does she communicate with other stars or with the galactic core? Magnificent, energetic movements of resonating radiance, electromagnetic resonances within the sun, produce vast electromagnetic flares, amazing patterns that appear as plasmodial mathematical shapes. How does this radiance communicate? Electromagnetic field theory, that foundation of our radio, television, and digital networks, points to the phenomena of resonance in the frequency domain as the basis of all communication theory. Resonance occurs when two systems vibrate at the same frequency; when they are thus "tuned into" one another, they then merge frequency vibrations into becoming a single resonant system, sharing information dynamically.

Human sensory systems also operate through resonance. Certain cells in the retina are resonant to a narrow band of frequencies in light. Certain cilia in the ears resonate with a narrow band of frequencies in air. Is it not likely that frequency resonance is involved in consciousness and communication with hidden dimensions of the metaverse? The essential technique of psychonautics is developing the ability to tune various regions of one's conscious mind to the higher frequencies beyond the normal operation of human consciousness in order to reach those "higher worlds" of awareness, higher bandwidths of conscious, communicable information.

My own first perceptions of the actual existence of these hidden dimensions occurred on the Fourth of July weekend shortly after my twenty-first birthday.

My First Experience of the Hidden Dimensions of Consciousness

At the end of my third year of university studies in mathematics and physics at Rice University in Houston, I was hired as a summer intern computer programmer at the Point Mugu Naval Station in Ventura,

California. That summer the unexpected revelation of "hidden" dimensions of awareness began around a small campfire on a beach where the Little Sur Creek flows into the Pacific. At around midnight on that clear moonless night, and not quite sure what to expect, I ingested three small yellow tablets of LSD-25.*

Given my university courses in electrical engineering, my understanding of the universe was deeply influenced by such things as frequency charts, Fourier and Laplace transforms, and electromagnetic theories that mathematically describe invisible energy waves. Suddenly, and without warning, all these dry electronic and mathematical paradigms became vividly alive for me, like live glowing wire frames overlaid and energized with direct visual, auditory, and tactile experiences of entities beyond my-*self*, immersing me in a formerly unknown and unexpected enormous sea of energy!

Astonished, I found myself floating somewhere within this vast ocean of energy and consciousness with an intensity of sensation I had seldom previously experienced, even during the deep and vivid dreams of REM sleep. My consciousness had been plunged into regions and depths of awareness that seemed to be lying just beyond the shores of my own familiar mind. At some point during this experience, I suddenly had the acute and actual realization that we (meaning every-*one* and every-*thing*) are *all* immersed in planetary and galactic fields of holographic awareness, swirling in and out of our normally limited islands of consciousness.

Several years later, I was given a book by my early mentor, Dr. John Cunningham Lilly (1915–2001), a physician, neuroscientist, and fellow psychonaut. In his book, I came across a description that resonated intensely with my own recollections of "what it was like" that first night under the influence of LSD. Here Lilly briefly describes his own initial experience with LSD while floating in body-temperature water in one of his carefully designed lightproof, soundproof isolation tanks:

---

*The dimensions beyond space and time are "hidden" to human perception much in the same way that the visual perception of space is hidden to a person born blind. It is the author's belief (and experience) that numerous sensory systems lie dormant or incipient in each human being, but that with the proper knowledge they can be activated; each individual can "turn on" various sensory systems that open awareness to this wider metaverse.

> I am a small point of consciousness in a vast domain beyond my understanding. Vast forces of the evolution of the stars are whipping me through colored streamers of light becoming matter, matter becoming light. The atoms are forming from light, light is forming from the atoms. A vast consciousness directs these huge transitions.
>
> With difficulty I maintain my identity, myself. The surrounding processes interpenetrate my being and threaten to disrupt my own integrity, my continuity in time. There is no time; this is an eternal place, with eternal processes generated by beings far greater than I. I become merely a small thought in that vast mind that is practically unaware of my existence. I am a small program in the huge cosmic computer. There is no existence, no being but *this* forever. There is no place to go back to. There is no future, no past, but *this*.[6]

My own experiences under the influence of the entheogen had opened my eyes to a completely new paradigm of what it might mean to exist, not only as a physical being on this planet, but also among the stars, as a "conscious being" in a metaverse of vast dimensions. Something had suddenly thrown wide open the shutters of my awareness to reveal a larger reality. No longer could I regard the real world in the relatively constricted ranges that I had inherited from previous generations, childhood perspectives from which I had unquestioningly adopted the boundaries and limitations of my existence. Everything—reality—had now become so much more alive than I had ever before imagined.

I was able to continue in my role as an intern engineering student that summer and to earn my degree the following year. But the intense psychonautic experience of those other worlds on the beach that night had ignited within me a flame of enthusiasm for exploring and understanding consciousness. For weeks I could not stop thinking about how I had somehow been plunged with eyes wide open into a vastly richer and wider range of experience than I had previously thought possible. Suddenly I had come to understand consciousness as a "thing" in its own right, as a research subject that could be studied and explored. No longer was my fascination focused exclusively upon quantum physics, circuit design, and laser communication theory; during the next few months, as I resumed my engineering classes, I found that I just could not stop thinking about the word *consciousness* and what it implied, both for myself and for the human race.

That initial contact with what clearly seemed to be new modes of consciousness opening to "higher worlds" now left me to wonder what an even deeper familiarity with these dimensions might mean for the direction of my life. It had become clear that exploring consciousness was now to be the highest priority among my lifetime goals, and that I would henceforth seek to understand consciousness by all means possible, both in my academic studies and research, and perhaps even more importantly, through following the psychological method of introspection as defined by William James,* as well as directly through personal introspective experiences. I realized that to do this, I would not only have to try to master the subjects of psychology, religion, philosophy, and science, but even more importantly, I would have to explore as first-person observer those vast and mysterious oceans of the higher worlds, the hidden dimensions. I would then have to do my best to integrate my understanding with my experiences and then to share what I had learned and experienced with other psychonauts, researchers, and the general public.

Shortly after returning to the university that fall, I realized that my new passion lay in a field that, as far as I knew at the time, had as yet no methodology nor even a name! I spent much effort searching academic catalogs for a program of consciousness studies, but found nothing even close. Now, many years later, I use the word that precisely encapsulates the subject that has fascinated me over the decades, and that word is *psychonautics*.† Yet, even today, fifty years later, there are relatively few

---

*In his twelve-hundred-page masterwork, *The of Principles of Psychology* (1890), William James (1842–1910), widely known as the father of American psychology, defined the *psychological method of introspection* as "looking into our own minds and reporting what we there discover" (185).

†*Psychonautics*: The methodology for describing, explaining, and mapping the subjective effects of altered states of consciousness induced by various means (e.g., psilocybin, cannabis, or other mind-altering substances, prayer, mantra, dreams, meditation, and a wide range of psychophysical exercises). The primary tool for gathering data is the individual human consciousness of the researcher in which one voluntarily immerses oneself into an altered state of consciousness in order to explore and later to record and analyze the accompanying experiences. The term's first published use in a scholarly context is attributed to ethnobotanist Jonathan Ott in 2001.

formal departments specifically dedicated to "Consciousness Studies" in universities around the world. Any study of consciousness is often embedded within the broader disciplines of psychology, neuroscience, philosophy, cognitive science, and even religious studies. While these fields frequently host research centers, institutes, or specialized programs focused on the study of consciousness, as a stand-alone department, it is quite rare to find a focus exploring consciousness as a science in any of these programs. And any direct, introspective exploration of consciousness continues to be found primarily in religious settings, ashrams, and mystery schools, not in academia.

What I see now is an unexpected positive aspect to the current chaos that seems to be accelerating daily in every region of our planet. It would appear that it does seem to be motivating a significant segment of young adults (and older adults) to turn inward, to become what I would term psychonauts in efforts to explore the "higher worlds" for the answers that traditional religions and philosophies clearly do not seem to be able to provide, even in the midst of so much environmental and social chaos.

A growing number of these modern psychonauts relate that during their excursions into the unknown they have encountered not only a wide variety of conscious entities but also vast networks of interlinked consciousness entities, hierarchies of such networks forming meta-galactic networks. It is as if conscious energies of the metaverse fill the dimensions beyond space and time much in the same pattern as stars and plasma fill our spacetime universe. What incredibly fantastic, beautiful, holonomic patterns fill the metaverse!

But beyond theory, and perhaps more importantly, also presented in this book are numerous approaches to how one's individual consciousness can be "tuned" to perceive the hidden worlds of the metaverse. Would an opening to these higher dimensions really be worth the effort and potential dangers? A resounding "yes" would be heard from shamans, mystics, and psychonauts willing to experience, to explore, and to map the hidden dimensions of the metaverse, those vast regions of consciousness that are on the verge of being opened to human consciousness.

Throughout history, many individuals have discovered that their brain-mind systems can be reprogrammed to act as powerful tools for accessing vast new dimensions of conscious experience. However, consistently practicing these techniques requires a certain level of motivation and a clear understanding of the goal. Those seeking an easy or quick solution may be disappointed to find that considerable effort is necessary to open the portal to higher dimensions.

This can be, at the outset, difficult to obtain without some clear understanding of the objectives and a map of the territory to be explored. At the very least, repeated studies have shown that these psychophysical exercises cultivate and strengthen one's ability to maintain a calm, tranquil, centered state of being in the midst of a world filled with anxiety, change, confusion, and ubiquitous dangers. But for the serious seeker, these psychophysical exercises will eventually open portals into oceans of consciousness that reach within at least seven dimensions hidden beyond our cosmos of space and time.

3

How I Came to Explore the Hidden Dimensions

This chapter provides my personal insights as to how I came to view mysticism and physics as compatible, complementary approaches to what we perceive as reality. My fascination with the possibilities for acquiring psychic abilities was kindled at the age of twelve, while spending many hours reading and rereading a book I had found in a mail order novelty catalog called *How to Hypnotize Your Friends*. The book begins with a description of the life and work of the Viennese physician Franz Anton Mesmer (1734–1815), known as the father of hypnotism and hypnotherapy, or as he termed the phenomenon, "animal magnetism."[1] Mesmer became famous as the term *mesmerize* was quickly embraced by medical doctors for much of the following century. Practitioners were often called "magnetizers," and even today it is widely used by alternative medicine practitioners.

According to the book, Dr. Mesmer's theory of animal magnetism is based upon a mysterious, omnipresent, and powerful living energy that exists everywhere in nature. According to his theory, this living magnetic energy is continually flowing between and within all things. The book claimed that a serious student could learn to sense, feel, and eventually to manipulate this magnetic energy, and thereafter one would be able to guide it as it flows through other people and animals.* At the end of the

---

*This is reminiscent of both Chinese acupuncture and Japanese Reiki, as well as the traditional "laying on of hands" found in various cultures.

book were listed a series of training exercises to be cultivated and practiced until the field of animal magnetism could be sensed and further developed. Through regular practice, the book assured the reader, one would eventually acquire the awareness and mental skills necessary to hypnotize others through direct control of this universal "animal magnetism."

To develop one's ability in this practice, a set of exercises were suggested to be repeated daily, preferably during the periods of sunset or sunrise, as outlined below.

"Animal Magnetism" Meditation

Step 1: Empty

One sits comfortably in a quiet darkened room lit only by a single candle flame, positioned at arm's length at eye level. While gazing steadily at the flame with eyelids half open, one should try to "let go of one's thoughts," or "empty one's mind."

The exercise requires making a sincere, continual effort to drop any word, phrase, memory, or unbidden thought the moment it begins to arise, to just cut it off, quickly letting it go, so as not to feed it the attentional energy by which it might otherwise grow and demand attention.

By continual practice of this "letting go" exercise, one will develop the skill of being able to detach from normal internal interruptions of the mind before they have the chance of acquiring momentum and gaining strength, demanding even more attention and blocking access to the more subtle "inner sources of power." This practice has been called "pulling weeds in the garden."

Step 2: Listen

Through this sustained practice, over a period of even five minutes, the normal chatter of the mind can be silenced and halted, allowing the remaining primary awareness to move on to focus within even deeper states of silence through the continual effort to simply "listen," while not becoming caught up in new thoughts, observations, or memories such as "how long have I been sitting here?"

Step 3: Sense

> Having developed some skill in silencing the mind, one can then move on to sense the energy of the candle flame in a new way, to sense where the image joins the eyes, and then to follow the direct tactile sense of the image flowing ever deeper into the brain itself through pathways of the inner sensorium.

For several weeks I practiced gazing at a candle flame in my bedroom at night, however my efforts came to an abrupt halt one night when my mother opened my bedroom door, saw the burning candle flame, and in no uncertain terms forbade me to ever again light a candle in my bedroom. Thus ended my candle exercises, but the flame of curiosity had been lit. What I had barely touched could not now be extinguished!

My Earliest Psychotropic Adventure

My second distinct effort to explore hidden dimensions of consciousness occurred during my senior year in high school. While doing research for a high school term paper, I found an essay by James written in 1882 on "The Subjective Effects of Nitrous Oxide," in which he described his experiments with laughing gas. James came away from the experience with a glowing description of nitrous oxide, stating that the effect had produced a kind of subjective rapture occasioned by the ability to make "the center and periphery of things seem to come together in one clearly unbroken whole." He claimed that while intoxicated by the gas his mind had been jolted into a new order of consciousness. I was intrigued, and particularly captivated with his assertion that inhaling nitrous oxide led him to experience what he termed "cosmic consciousness." Of course. I immediately knew that I *too* wanted to experience directly what James had described.

With my very rudimentary high school chemistry knowledge and an enormous amount of hubris, I set out to manufacture laughing gas. As student president of the chemistry club, I had been given free access to a small lab in the back room of the chemistry classroom. I soon came up with what I thought to be a simple chemical equation to create nitrogen oxide, and at last managed to generate a beautiful shade of brown hued gas in a large Erlenmeyer flask. After briefly cooling the beaker of gas in a bucket of ice water, I opened the stopper and inhaled a deep lungful of the gas and began to hold my breath to experience the effect. As I breathed it in, the gas was tasteless and easy to inhale. Soon after came the surprise. After about thirty seconds, I began to exhale, or at least tried to exhale, but something was wrong. It was as if the gas wouldn't leave my lungs. Somehow, I was not able to inhale a fresh lungful of oxygen, and I began to panic. I don't remember passing out, but my next conscious memory was looking up from the cold floor of the lab room into the eyes of my chemistry teacher, leaning over me and shaking me by the shoulders, asking loudly what had happened and if I were all right. Coughing and choking, I soon recovered, but I have no recollection of what excuse I gave Mrs. Opp, my teacher and mentor. Definitely my first attempt (and almost my last) to experience a chemically induced widening opening of consciousness was an abject failure.

But my educational trajectory was still firmly planted in the direction of hard sciences, primarily physics. The following summer I was accepted into a summer intern program at the Naval Weapons Research Lab at Fort Belvoir, Virginia, not far from my family home, under a National Science Foundation program to which I had applied. Nervous at first, as always, I soon found to my surprise that my research scientist mentors, rather than being strict, critical, and demanding as I had feared, were relaxed and friendly, and all seemed to have the same proclivity for practical jokes. They found special glee in tormenting a recent physics graduate from Caltech, Joe Taylor, who was working on the electrical properties of the common earthworm. One day Dr. Joebstl, the electron microscopist who spent most days tinkering with electronics, dipped one of the Joe's worms into a vat of liquid nitrogen, and then, when Joe was out for lunch, dropped it onto the workbench where it immediately shattered into hundreds of worm shards.

The research scientists were mostly working on the study of an explosive class of molecules called inorganic azides that the navy wanted to adapt to high explosive weapons. However, since my contributions, as a high school student for which they were responsible, could at the best be minimal or worse in working with dangerous explosive materials, the scientists encouraged me to work instead on a science fair project that I had mentioned to them. With their help and their state-of-the-art machine shop, they helped me to design and construct (to my utter amazement) a small "linear ion accelerator" for my science fair project.

The final device far surpassed my expectations, visually out-classing the best futuristic technologies I had seen in science fiction films. My accelerator consisted of a 36-inch long, 6-inch diameter clear vertical Pyrex glass tube mounted on a machined brass base with a 12-inch hollow aluminum ball on top, and a heated tungsten filament. To the brass base we attached a small vacuum pump to draw out the air from the glass tube. Adjacent to the glass accelerator tube we placed a large Van de Graaff generator with a second 12-inch hollow aluminum ball on top. The two aluminum spheres were placed in contact with one another, and so when I fired up the motor, upward of 500,000 volts of electrons would soon gather on the aluminum balls, emanating high voltage sparks over 30 inches in length with a satisfying characteristic lightning snapping sound. Around the glass accelerator tube I had coiled several rings of copper wire, and connected these to a surplus army communications transmitter from World War II.

After creating a good vacuum in the Pyrex tube, I turned on the Van de Graaff generator and heated up the filament until it began glowing brightly at the bottom of the accelerator tube as it gave off electrons. Within the vacuum inside the glass Pyrex tube, the electrons given off by the electrode began to flow upward through the vacuum toward the charged Van de Graaff sphere. What appeared to be a twisting aurora borealis began to glow and dance: suddenly an amazing brightly colored, dancing plasma appeared, much like images I had seen of the aurora borealis over the far northern skies! I thought it was visually stunning, and so did the Northern Virginia Science Fair judges that following November. Years later I discovered that David Bohm, working as a graduate student under Robert Oppenheimer at UC Berkeley, had experienced a similar encounter while viewing a plasma experiment. It seemed to him as if the glowing plasma were somehow a living thing, and the phenomenon so intrigued him that he began to develop his doctoral dissertation on the physics of plasma, a paper that became so important to the development of the first atomic bomb that it was marked Classified, never to be published.

I was awarded first prize at the science fair for my experiment, and this, I believe, went far in helping my being accepted for admission into the Massachusetts Institute of Technology. However, I also had been offered a full scholarship to study physics at the Rice Institute, and as my family had just relocated to Texas, I decided to enter what is now Rice University in Houston, Texas.

The Summer of Love (1967)

In my third year as a physics student at Rice, I married a fellow student whom I had been dating for several years, a painter in the fine arts program at the University of Texas in Austin. It was during our first summer together as married students that my lifelong passion for exploring the science of consciousness was kindled by my initial experience with LSD.

During the 1967 July Fourth weekend we drove north along the coastal highway with our friend Roscoe to a beautiful region of the Big Sur coast, halfway between Los Angeles and San Francisco. Roscoe had with him several little yellow tablets of what he said were "Owsley Sunshine," high quality LSD produced by the clandestine chemist, Owsley Stanley. Our friend insisted that we would have a fantastic "trip" and maybe even see God. Being a young scientist brought up by Catholic parents, I was intrigued by a recent *Life* magazine cover article about people who had taken LSD reporting that they had "seen or communicated with God." Out of both scientific and religious curiosity, I wanted to experience a full dose of LSD and perhaps even "commune with God."

In the early evening, we parked by Highway 1 about twenty miles south of Big Sur just as the sun was setting in orange, pink, and blue. We made our way down the sand dunes to the beach far below and soon found driftwood to build a small fire, sheltered by sand dunes leading to the beach. In the early dark of night, at about 9:00 p.m., we swallowed our yellow tablets. Young and foolish as I was, and wanting to ensure the effect would be perceivable, I took three of the tablets. I was later told that each Owlsey tablet contained 250 µg of pure lysergic acid diethylamide, a fairly strong dose by itself, so that I had ingested a rather large dose (even by 1967 standards) of 750 µg.

Shortly after, while staring into the flames of the small flickering fire, as I entered the mind-altering dimensions of experience catalyzed by this entheogen, my worldviews of both science and religion were forever changed. Is it even possible to put into words what I experienced that night? I truly believe that those eight to ten hours, spent on a chilly central California beach, changed the trajectory of my life. My future efforts, both academically and experientially, were henceforth to explore the amazing dimensions and networks of consciousness to which I had been linked during that night under the stars.

Afterward and for some time, all that I could remember of this intense episode were the initial physical effects, the feeling of electrical currents flowing, first in my gums and then distinct electrical, lightning-like snapping sounds and sensations somewhere inside my cranium and down through the back of my neck. Then at some point these sensations culminated with an enormous and abrupt shift in my awareness, an indescribable change in sensation and perspective that Terence McKenna often referred to as "the rupture of plane," a phrase which I was later surprised to discover had also been used by the great philosopher of religion, Mircea Eliade used in describing the goal of deep contemplative practice, the Sanskrit term *samādhi*: "In *samādhi* there is the 'rupture of plane' that Indians seeks to realize—the paradoxical passage from Knowing to Being."[2]

Trying to remember the ensuing experiences that evening was as impossible as recalling a vastly complex dream sequence, but I took away a deep conviction that what I had inadvertently stumbled across "there," wherever I had been, and however I had arrived "there," had shown me an ocean of knowledge and experience that could be tuned into, but one that normally was so filtered out from awareness as to be completely concealed from everyday waking consciousness.

During that long and amazing night under the stars on that beach at the mouth of Little Sur Creek, at the peak of my educational immersion in mathematics and electromagnetic science, I had come to realize that something called consciousness, a subject seldom mentioned by my university professors, had suddenly become more important than any other of the subjects I had been studying. It was there, under the immensity of stars, by the shore of the vast dark Pacific, that I had experienced living mysteries that lie far beyond those of science or religion. Henceforth I would be working toward exploring, charting, and understanding the oceans of consciousness that I had experienced and within which we all exist.

Several years later, I found my experience described almost identically in the passage I mentioned in chapter 2 by John Lilly describing his own first experience with LSD-25, in a lightproof, soundproof, isolation tank. His words were so powerful, let's review them again:

> I am a small point of consciousness in a vast domain beyond my understanding. Vast forces of the evolution of the stars are whipping me through colored streamers of light becoming matter, matter becoming light. The atoms are forming from light, light is forming from the atoms. A vast consciousness directs these huge transitions.
>
> With difficulty I maintain my identity, my self. The surrounding processes interpenetrate my being and threaten to disrupt my own integrity, my continuity in time. There is no time; this is an eternal place, with eternal processes generated by beings far greater than I. I become merely a small thought in that vast mind that is practically unaware of my existence. I am a small program in the huge cosmic computer. There is no existence, no being but *this* forever. There is no place to go back to. There is no future, no past, but *this*.[3]

No longer was my fascination focused exclusively upon radio and laser communication theory and an engineering career; suddenly I discovered that I could not stop thinking about the word *consciousness*, and what it might imply, and how it might operate, and how inner space might be explored, somehow. Thus began my lifelong passion for exploring, mapping, and trying to understand the dynamics and architecture of consciousness itself.

It was the Summer of Love, and we quit our summer jobs at Point Mugu a month early to travel up the California coast on a pilgrimage to San Francisco and Haight Ashbury, the mecca of hippiedom. Earlier that summer I had acquired a book in Los Angeles by academic hippie-heroes of the times, *The Psychedelic Experience: A Manual Based on the Tibetan Book of the Dead* compiled by Leary, Metzner, and Alpert. Always impressed by academic credentials, I carefully read the book and found it unlike any subject I had ever previously read, whether science or fiction. It dealt with contemplative practices to accompany altered states of consciousness as one approaches the death state, and apparently it was a translation, in modern psychological (and somewhat hippie) terms of an ancient Tibetan Buddhist text that was to be read aloud during the passage of a person from life to death. Leary, Alpert, and Metzner had reinterpreted the text to include its application to changes in consciousness brought about by "ego death" of an individual going through stages of the psychedelic experience. The goal was to let go of ego "hang-ups" and release one's old identity/identities to facilitate going deeper toward "the Clear Light" of universal cosmic consciousness. It was a most interesting book for me at the time.

I was intrigued to learn of the various experiential states that could be reached during the death transition, or through taking psychedelic drugs, or through various forms of contemplative practice. It was the idea of traditional contemplative practices that most intrigued me, and I felt drawn to learn forms of yoga and alchemical techniques (often discussed by Carl Jung) that might lead to entry into these altered states within the hidden dimensions.

A UFO at Hamilton Pool

Back in Texas for my final year in electrical engineering, my grades dropped, likely because my overriding motivation had taken a different direction. Every few weekends we would go out into the Texas Hill Country at night and experience new dimensions of conscious reality through experimenting with peyote, psilocybin mushrooms, mescaline, or LSD. In Austin, peyote was not yet illegal and could be obtained at a local garden shop. These were the early hippie days, and it seemed as if a vast new wave of psychotropic exploration had begun. I was now twenty-one, and this inner world of incredibly energetic dimensions was new and full of amazing potential for discovery and experiential research, full of unimaginable entities and realms both visual, audible, and empathic.

And yet even more was to come to shift my understanding of the world, and the following spring I again experienced a major paradigm shift. It came with the appearance of a strange sphere of living light that my friends and I encountered in the Texas Hill Country near Austin. This occurred near Hamilton Pool, a small grotto formed by a tributary of the Pedernales River about thirty miles west of Austin, along an ancient crack in the earth called the Balcones Fault. Fed by a small waterfall, and about three hundred yards across, the pool was deep and dark, almost a perfect circle. It was around midnight, and we had come to rest and swim after a week of final exams. The night was quiet except for the chirping of insects, broken occasionally by the sound of croaking bullfrogs. A thick pale ceiling of moon-backlit clouds concealed the stars. We arrived around 9:00 p.m. from Austin, having succumbed to a late evening urge to go swimming in the countryside to celebrate the end of the semester, and at around 10:00 p.m. we had taken what we considered to be a relatively small dose of LSD, each swallowing half of a small yellow Owsley acid tablet. During the first half hour we sat at the edge of the pool, leaning our backs against a large boulder, talking quietly, often lapsing into silence. Soon I began to feel the physical effects of the entheogen as a familiar buzzing in my gums, accompanied with brief visible flashes of phosphene light in my peripheral vision.

The first sign of something externally out of the ordinary began with an abrupt cessation of the crickets' chirping, which seemed to cause an amplification in the sound of the waterfall on the other side of the pool, splashing down from an overhanging rim directly across from where we sat. Soon even the intermittent bullfrog croaking stopped, and the dark silence was broken only by the regular sound of water falling into the pool.

After what seemed to us an interminable time even our thoughts faded away. A barely perceptible glow seemed to flicker far down the creek that drained Hamilton Pool into the Pedernales River. The light seemed to be growing and fading in a slowly pulsating cycle. The light, which at first seemed far away down the creek, moved steadily closer, sometimes fading, but then growing brighter as it pulsed in a slow regular rhythm, continually moving toward us, intermittently obscured in the distance by intervening trees of the forest-lined creek. The glowing light at last emerged from the greater darkness of the night foliage, clearly visible above the bed of the stream, moving ever closer to the pool. Slowly, almost majestically, the flickering sphere, about the size of a large beach ball, circled the rim of the overhanging cliff that surrounded two-thirds of the grotto in front of us. When the apparition reached the falls, it paused for what seemed to be a long time, then very slowly dipped downward several feet, in front of the falling creek water, and began a strange pattern of bobbing and weaving in a rhythmical pattern in front of the falling water, as if it were some gigantic moth in the night. Eventually it stopped, becoming motionless. Soon it resumed its circuitous path along the sheltering overhang of the grotto rim, moving about ten feet above the shore, ever so steadily moving toward our side of the pool, as we watched in awe, incredulity, and some measure of fear.

At last, it came close to the rock by the shore where we sat, completely still, in suspended judgment. It paused for what seemed to be an eternity, hovering ten to twenty feet above our heads. To me it appeared like a galaxy of tiny stars coalescing within a spherical space the size of a beach ball, yet each point would move in small linear traces, as if there were a thousand points of light tracing straight-line paths in seemingly random patterns, yet all circumscribed within what seemed to be a perfectly spherical boundary about a meter in diameter.

For what seemed to be an eternity it hovered above us, completely still other than the myriads of tiny wire frame light paths within its bounding sphere. At last, it resumed its circumnavigation of the grotto, moving once again toward the creek draining the pool, then following the creek, bobbing, and weaving on its way back into the forest and the swampy land around the creek from which it had emerged.

What I felt and perceived during that time in the presence of, and then beneath, this apparition I can scarcely remember, let alone begin to describe, but the long-lasting impact on my psyche and understanding of the world is still with me, decades later. I recall hearing peculiar snapping electric-like sounds within my head, and an intensity of feelings arose in me during the episode that I had never before experienced. I did know with absolute certainty that the apparition had radically changed my view of the natural world, my life, and my previous unquestioning view that science had already mapped and understood most of the world. Here, thirty miles or more from the nearest "civilization," in the silence of a simple Texas countryside, a major unknown had revealed itself.

That this apparition, this being or entity of light, possessed awareness, consciousness, that it perceived us and paused above us, there could be no doubt. For days it was all we could talk about, and we discussed the apparition at length. We wondered if it should be reported, and if so, to whom. Was it a UFO? All reports I had ever read of UFO sightings had assumed

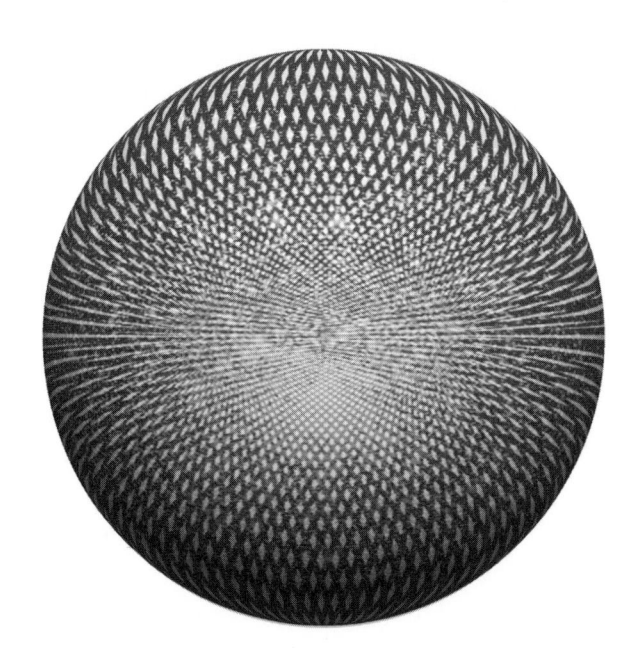

Fig. 3.1.
Moiré patterns in a spherical volume.

the apparitions to be machines, mechanical vehicles of some type. But it seemed clear to the three of us that this entity had been a living being of the forest, somehow, moving with intention in the way a living creature might move. It was alive, clearly, though not made of any identifiable material or substance; it had appeared as a transparent sphere, manifesting glowing electric sparks and lines in visibly changing swirls, as might be seen in holographic moiré patterns (fig. 3.1).

We later agreed that as the sphere hovered above us the impression was one of benevolent curiosity. Could this have been some kind of machine? Its movements did not seem machinelike, but gave the impression of intelligence, particularly in the lovely "dance" it made in front of the waterfall, and later as it hovered directly above us, twinkling and glowing in pastel aurora-like colors. The sphere had seemed to be considering us, observing us intently. We had all been both terrified and amazed, like deer caught in car headlights, we had frozen and could neither move nor think nor react.

Many years later, as I continued what had become my lifetime study of consciousness in a search for corroborating material, I was fortunately able to find descriptions of similar encounters in the writings of both John Blofeld (1913–1987) and Aleister Crowley (1875–1947).[4]

Born in London in 1913, Blofeld was fascinated by Buddhism and contemplation from an early age. He traveled to China in 1937, spending the next decade visiting remote monasteries and sacred mountains in China, Mongolia, and Tibet. In *The Wheel of Life: The Autobiography of a Western Buddhist,* Blofeld relates that he once visited the Temple of Wutai Shan on a mountain of the same name in northern China, sacred to the Bodhisattva of Wisdom, Wénshū (文殊 in Chinese). Here Blofeld relates his experience one night on the top floor of a three-story meditation tower constructed on the slopes of the holy mountain of Wutai Shan:

> The ascent to the door of the tower occupied less than a minute. As each one entered the little room and came face to face with the window beyond, he gave a shout of surprise, as though all our hours of talk had not sufficiently prepared us for what we now saw. There in the great open spaces beyond the window, apparently not more than one or two hundred yards away, innumerable balls of fire floated majestically past. We could not judge their size, for nobody knew how far away they were [. . .] Fluffy balls of orange colored fire, moving through space, unhurried and majestic—truly a fitting manifestation of divinity![5]

The temple was renowned for the fact that (according to the Buddhist monks there) bodhisattvas could often be seen floating down from the mountaintop in the form of spheres of light, and special meditation platforms had been constructed so that the balls could be observed by contemplatives. Blofeld continues in an attempt to categorize what he and his companions had seen that night:

> I do not know if this extraordinary sight has ever been accounted for "scientifically" and I am not much interested in such explanations. It is far lovelier to think of them as divine manifestations, however prosaic their real nature may be. But is it prosaic? Marsh gas, you say? Marsh gas right out in space, a thousand or more feet above the nearest horizontal surface and some hundreds of feet from the vertical surface of a cold, rocky mountain innocent of water? Surely not.[6]

Aleister Crowley, the British painter, theosophist, and mountaineer recorded in his biography the details of a similar experience of a seemingly living sphere of light. He describes a sphere of light that entered his hut by a lake in Scotland during a violent thunderstorm. He writes that at first he thought it must be St. Elmo's fire, balls of light often seen hovering about the top masts of sailing ships during thunderstorms at sea, but the one that appeared to have entered his cottage during a storm seemed clearly to be alive as it moved in a stately procession around the room, as if studying the environment, then dashing straight toward Crowley's face, only to stop suddenly, and after what seemed to be an interminable period, finally moving slowly back out of the hut and vanishing into the storm.[7]

In the light of this experience, my entire view of the world went through an upheaval. Though I was a freshly trained engineer and had been inculcated into the mysteries of mathematics, physics, and the paradigms of an empirical materialism for the previous three and a half years, yet I was young. My ideas were too fresh for them to have settled into an inflexibly defensive posture and thus the experience opened a large breach in my prior confidence that the sciences were close to explaining everything in the universe. A new world had revealed itself.

I recalled the vivid impressions of a lecture I attended, given by the Nobel Prize-winner Richard Feynman, during which he challenged his students to close their eyes and to use our imaginations to visualize what the electromagnetic energies of the planet and universe would look like if we could view them as they pass through the very lecture hall merging and patterning one another in geometric swirls and rainbow-hued colored spectrums as they passed through our bodies and minds, linking us with other galaxies and to one other. As remarkable as Feynman's thought experiment was, it still maintained the materialistic science vision of a sterile universe, a mechanical matrix of beautiful yet nonliving energies. Suddenly the light in the forest resonated with my memory of Feynman's lecture and catalyzed my own awareness into the realization that these energies must be alive, that these energies exhibit consciousness in some mysterious way in which my previously limited, scientifically and logically conditioned framework of understanding would never have allowed.

That fall semester I returned to my studies of electricity, electronics, and physics with a new perspective but with new enthusiasm. I found that I was beginning to understand electronics in ways that I could not easily share with my classmates who all seemed to be mindlessly adopting the "Universe is (mostly) dead matter and random motion, but if we jiggle things right, we can make useful devices" paradigm. I began to experience a growing awareness that the very electricity flowing within the building walls must be alive in some authentic, broader than biological way, that the light energy flowing from light bulbs, the frequencies carrying radio and television signals, the entire electromagnetic spectrum surrounding all of us might just consist of living fields, in some dimension of consciousness, and that electrophysics itself could, and perhaps should, be a subfield of biology. Most significant of all was a growing awareness of the complex flows of radiant energy emanating from and coursing throughout my own physical body and between and among those around me.

On weekend evenings we would regularly drive out into the countryside to explore the incredible new dimensions of reality revealed to us with the assistance of various entheogens. At that time, it was possible to purchase live peyote cacti at a local nursery in the outskirts of Austin, and one of our friends became accomplished at boiling the peyote buds for five hours to produce mescaline sulfate crystals. The drug was made famous through the writing of Aldous Huxley,* who took mescaline for the first time in the spring of 1953, and the following year published an account of his experience in *The Doors of Perception*. And then there was that new and quite mysterious book based upon *The Tibetan Book of the Dead* that was assembled by Richard Alpert, Ralph Metzner, and Timothy Leary, which opens with the following:

> A psychedelic experience is a journey to new realms of consciousness. The scope and content of the experience is limitless, but its characteristic features are the transcendence of verbal concepts, of spacetime dimensions, and of the ego or identity. Such experiences of enlarged consciousness can occur in a variety of ways: Sensory deprivation, yoga exercises, disciplined meditation, religious or aesthetic ecstasies, or spontaneously. More recently they have become available to anyone through the ingestion of psychedelic drugs such as LSD, psilocybin,

---

*The philosopher Aldous Huxley can surely be considered a psychonaut: on his deathbed and unable to speak owing to advanced laryngeal cancer, Huxley made a written request to his wife Laura for "LSD, 100 mg, intramuscular." She obliged with an injection at 11:20 a.m. and a second dose an hour later. Huxley left his body six hours later at 5:20 p.m.

mescaline, DMT, etc. Of course, the drug does not produce the transcendent experience. It merely acts as a chemical key—it opens the mind, frees the nervous system of its ordinary patterns and structures.[8]

During that time it was clear to me that my own mind was definitely opening to these vast new realms of consciousness!

In June of 1969, after graduating with a degree in electrical engineering, I flew to New York City, where I continued my reading about contemplative practices and altered states of consciousness. I continued to practice meditation, but my career goals had vanished and I had no idea of what the future might hold. I had been reading a lot of Paul Bowles novels and short stories about Morocco and yearned to break out of my somewhat depressing, aimless situation. Next door to our building was a huge Maritime Building and I began to wonder what it might be like if I could become a sailor and travel to far-off ports, perhaps find Paul Bowles's North Africa, and even perhaps sail to India.

I think it was my passion for reading that helped maintain my sanity during this stressful, uncertain, emotional period of my life. I had discovered a public library a few blocks away. At the time I was searching through subject areas that dealt largely with psychology and science, hoping to find material relevant to consciousness and in particular to "out of the ordinary" experiences. One Saturday morning I noticed a book someone had left on a reading table, and I opened it and began reading. The title of the book was *In Search of the Miraculous: Fragments of an Unknown Teaching,* and I was surprised to learn that the author had been a mathematician, a Russian named Pyotr Demianovich Ouspensky (1878–1947). Having always been impressed by technically credentialed writers, I checked the book out of the library and, with a mixture of awe and reverence, spent the rest of the weekend reading Ouspensky's book. Here I found a deep, articulate discussion of experiences and concepts that directly addressed the subject of consciousness that now so fascinated me. Up until then I had assumed that all books fell into one of two categories, technical nonfiction and fiction. Yet here was a book, written by a professional mathematician, describing his experiences during his three-year exploration of consciousness and search for "the miraculous" under the guidance of his teacher, a Greek-Armenian Orthodox Christian mystic named George Ivanovich Gurdjieff

(1867–1949). Reading the first page, I was immediately captivated, finding in Ouspensky's opening passages, word for word, a description perfectly mirroring my own search. Ouspensky writes here of his search for the miraculous:

> I had said that I was going to "seek the miraculous." The "miraculous" is very difficult to define. But for me this word had a quite definite meaning. I had come to the conclusion a long time ago that there was no escape from the labyrinth of contradictions in which we live except by an entirely new road, unlike anything hitherto known or used by us. But where this new or forgotten road began I was unable to say. I already knew then as an undoubted fact that beyond the thin film of false reality there existed another reality from which, for some reason, something separated us. The "miraculous" was a penetration into this unknown reality. And it seemed to me that the way to the unknown could be found in the East.[9]

Of course, for me, the meaning of "miraculous" had become "consciousness." Ouspensky's book opened my eyes to an entirely new category of writing that I had not previously imagined existed, and suddenly I realized that what I had been seeking in science and psychology might be found in the "esoteric" subject areas of mysticism, occultism, theosophy, and Asian religions and philosophies; all of these were topics I had never before been introduced to in high school or university courses, and had until recently been completely ignored by the scientifically trained communities (notwithstanding the recent interests of the psychologists Timothy Leary and Richard Alpert).

In addition to suddenly "discovering" an entirely new range of reading material to be explored, I soon acquired the habit of attending various lectures and workshops dealing with meditation and the esoteric; New York in the early 1970s had become a confluence for New Age teachers and authorities on esoteric contemplative practices, and over the next several years I was able to study with a number of seemingly authentic teachers, including Swami Satchidananda Saraswati, Chögyam Trungpa Rinpoche, John Lilly, and Alan Watts. But during my five years in New York I had no idea that there might also be rich mystical traditions and teachings within my own Christian roots, and so my focus was upon absorbing theory and techniques regarding mysticism and psychonautics from Buddhist, Taoist, Hindu, and Native American traditions.

Perhaps it was synchronicity, but I was fortunate to be living on East Sixth Street, only three blocks from the esoteric bookstore run by the publisher Samuel Weiser.* Weiser's Bookstore offered many hundreds of books on such topics as Christian mysticism, Vedanta, Theosophy, yoga, Buddhism, Sufism, medieval metaphysics, magic, shamanistic traditions, and a wide variety of esoteric religious teachings from many cultures; there, I spent many hours after work and on Saturday mornings browsing the shelves in my search for coherent theories of consciousness that might be compatible with what I knew of radio theory, physics, and quantum mechanics.

Fasting, Yoga, New York City

The experiences of Little Sur Creek in 1967 and the encounter at Hamilton Pool in 1968 greatly reinforced my motivation for regular practice of meditation, hatha yoga, and fasting exercises. I stopped eating meat entirely, and went on frequent water fasts, losing thirty pounds during the spring of 1970. Initially, I would first do physical exercises and hatha yoga, followed by what seemed to be a surprisingly long ten minutes (using a timer) trying to meditate, watching my breathing and trying to quiet my mind, trying to free it from the rising of random memories, interior verbalization, and other impulsive cognitive distractions. While at the beginning the ten minutes was an arduous exercise, over the next few months I was able to increase the time in five minute increments.

I became interested in all forms of meditation and began attending lectures and weekend workshops given by Indian and Tibetan teachers. Soon I began following a Tibetan teacher named *Chögyam* Trungpa who was reported to be a holy *tulku* or reincarnation of a former Tibetan saint. Trungpa was unusually modern in his presentation. He had escaped from Tibet during the Chinese invasion and had learned

---

*Samuel Weiser's Bookstore, opened in 1926, was known as the oldest and one of the most famous metaphysical bookstores in the United States. It is now known as Weiser's Antiquitarian Books, and continues to operate a thriving business in esoteric writings.

English while living in England before coming to New York. Trungpa was known for his unorthodox approach, which included challenging traditional religious norms, breaking cultural taboos, and engaging in behavior that some saw as reckless or even scandalous.

His approach earned him the nickname "the crazy wisdom master," a term derived from the Tibetan concept of *yeshe chölwa* or "crazy wisdom." This concept refers to enlightened behavior that transcends conventional moral standards, often appearing eccentric or unpredictable to those who do not understand the underlying spiritual purpose. I found his lectures and instruction in New York fascinating and soon made my way for retreats to his Vermont meditation center called "Tail of the Tiger" where I would watch Trungpa drink five or six large cans of a strong malt liquor sitting beside his throne on a silver tray while he gave his dharma talks. Occasionally a new visitor, clearly disturbed by watching a spiritual master drink heavily, would bring up the subject. Trungpa would then focus on the theme of "hangups" and why one should confront them directly. Responding honestly to my own "hangup reaction," I soon gravitated away from Trungpa's group darshans, but continued to study Tibetan mysticism and contemplative techniques through other means. During my second year in New York, I came across an advertisement for a lecture to be given by Dr. John Lilly (1915–2001) on *Programming and Metaprogramming the Human Bio-Computer*, the title of a book he was about to publish.

Intrigued by the topic, I attended the lecture, which was presented to an audience of twenty or so in a small hotel room near Carnegie Hall on 56th street. His talk made an enormous impression on me, perhaps because I had been so recently immersed in studying computer languages, electromagnetic theories, and exploring firsthand the effects of psychotropic substances upon my own consciousness. His talk involved all three topics, which were central to several years of research that he had conducted in "interspecies communication." His theory was that the brain operates as a biocomputer, and that one can learn to program and reprogram the operations of the brain, and thus consciousness itself, through the practice of silent contemplative techniques in a soundproof, lightproof environment. In the late 1950s Lilly had established a research center called the Communications Research Institute on St. Thomas in the U.S. Virgin Islands. There he conducted a series of experiments, condoned and financially supported by the National Institute of Mental Health. The experiments involved administering clinically pure LSD-25 to dolphins and their trainers and recording their audible and subaudible interactions as they floated in large indoor pools of water.

Lilly's extensive knowledge of electronics and software made his ideas all the more fascinating, even more so when I discovered that, like myself, he had also become a licensed HAM radio operator at an early age. His obvious enthusiasm for exploring the psyche in every way possible, through science as well as by direct introspective experience, was contagious. I was particularly encouraged by his assertion that individuals with backgrounds in physics and electronics, having developed the capacity to focus for extended periods upon abstract concepts, would find considerable success in applying the esoteric techniques of contemplation. In the future, he said, scientists will fill the ranks of a new generation of mystics. It was under Lilly's guidance that I constructed a small meditation chamber in my own loft on the Lower East Side and began a lifelong practice of daily contemplative practice in the quietest, darkest location I could find.

First Experience of Inner Sounds

During my mid-twenties I was living in a two-room flat on the fifth floor of a five-story walk-up. Though it was a challenge when carrying heavy groceries, my apartment on the fifth floor had the advantage of lowering much of the noise of the city neighborhood, and my inner room was wonderfully quiet, especially at night. Late one evening I was in that inner room doing my usual stretching exercises, trying to maintain a shoulder stand posture (*sarvāṅgāsana*) for ten minutes as part of my hatha yoga practice. Part of the exercise was to move into the pose, then to become as quiet as possible, practicing internal silence. As during formal sitting meditation, this hatha yoga meditation required an effort to attenuate every thought that might arise, to detach from and not follow memories as they began to form, nor to allow any inner dialogue to resume streaming. The goal was to open up the bandwidth of awareness and to remain receptive, just listening, while relaxing into the yoga pose.

Suddenly, out of the silence, I heard a singular loud, high-pitched tone that seemed to be emanating from somewhere within the right-hand region of my cranium. I soon noticed that as I focused my awareness on the sound it seemed to coalesce into a distinct single point while substantially increasing in volume! I was on the verge of panic, fearing that I might be experiencing a brain aneurysm in progress. But as I soon discovered that by maintaining my focus, I was able to coax the sound into growing even louder and more distinct, my fear was transformed into awe at this audible tone coming from within. Even more strange was that accompanying the sound sensation was a sensation of "touch" detectible within this tiny region located somewhere within the upper right-hand quadrant of my brain.

Then things became even more strange. After noticing the initial "bright" sound, additional "points" of sound at distinctly different pitches began to rise into awareness *in other locations in my cranium.* I gently lowered myself from my shoulder-stand position and, ending my yoga for the night, lay down under a blanket in the dark. For many hours that night I could not sleep, totally fascinated by focusing upon and listening to the sounds that would variously increase in volume according to the degree that I would be able to direct my attention toward them. I noticed, however, that as soon as I would begin consciously thinking "about them" or "thinking in words," letting my attention begin to stray, they would subside, and contact would be lost. I quickly learned that by gently dropping my train of thought, which seemed so insistent on thinking, classifying, and so on, I was able once more to enter the silence, and the tiny sounds would suddenly peek out of the silence again, and increase in volume in what was clearly a feedback loop, a sort of reverberation responding to my search. The tones were quite pure and high-pitched, and I suppose most people would classify them as a "ringing in the ears." Several months later I discovered the term "tinnitus," which was defined by medical science as any perceived sound not brought in by the ear canal. Since perception of these sounds seemed to bother people, doctors decided that it must be a disease of the auditory nervous system with an unknown, yet to be determined source.

Nevertheless, by now being quite serious in my efforts to explore the phenomenon of "consciousness" by any means possible, catalyzed by the unknown phenomena I was now experiencing during meditation and encouraged by my previous encounters with the mysteries at night in Big

Sur and Hamilton Pool, I found myself fascinated by what was happening in my body. I found that by trying to ignore a particularly dominant bright sound and shifting my focus to a fainter, more obscure sound ("further away from" or "behind" the first), the second sound would immediately grow louder in volume and become easier to discern. Here was direct cause and effect, the direct experience of being able to guide and control an energy phenomenon of consciousness within my own brain region. All that night I lay awake in the dark, moving from sound to sound within my head, as each would rise and fall, almost as if each had an independent volition of its own. I experienced strong emotional oscillations between exaltation verging on disbelief, and fear that I might be damaging my neuronal centers, perhaps even triggering some variety of brain-damaging hemorrhage.

As an electrical engineer, I had often listened to various single sinusoidal tones generated by equipment in laboratory sessions; yet this was not a single tone, but a confluence of tones faintly making up a background of the perceived, sensed audio range, reminiscent of the pervasive "peeper" sounds I had heard in the dark night at Hamilton Pool. From time to time a specific new tone would arise with exponential sharpness high above the background level, to become a bright point, like a beacon that, if I were able to sustain focus for a few moments, would become markedly louder with an accompanying intense tactile sensation.

During what seemed a very long night my body grew hot and sweated profusely, soaking the sheets in what I assumed might be a fever caused by whatever was happening in my brain. I went through what seemed to be a long period of deep fear, suspecting that I was somehow damaging my nervous system.

Sometime in the early morning hours I fell asleep. When I awoke it was with great relief to find that my mind seemed to be back to normal, having returned to its familiar mode of verbalized thoughts chatting away merrily once more. However, I now lived with these new memories and the realization that something singularly strange had occurred, something I had never been prepared for and which I had never previously encountered in books nor in life's experiences.

I continued to practice hatha yoga but spent increasingly long periods in silent meditation, finding that, now, I was able to fairly easily contact these resonant inner sounds. I began the practice of focusing upon them while falling asleep and found that when I would begin to awaken from a dream in the middle of the night, I was able to quickly reenter the dream world by following these mysterious bright inner sounds. Over the next several years this process of concentrating and visualizing within areas of my body while focusing on the sound tones as they would arise became a main source of meditative practice for me, and the inner tones grew ever more richly complex and often markedly louder in volume, and began to produce distinct tactile sensations of a flowing nature, unlike the sensations felt in the external senses of touch, vision, taste, and hearing. My training in physics and electrical engineering led me to believe that these internal sounds were sine waves, not some sort of random noise. I noticed as well the tones manifested around specific, fundamental frequencies, grouped within narrow spectrums. For a time, I conjectured that they stem from mechanical resonances within the physical structures of my inner ear. At work I began to experience, with great surprise, one of the high-pitched sounds flare up in my cranium whenever I approached certain electronic equipment, computer screens, or even certain vending machines. At such moments I found myself internally verbalizing, with some humor, "Incoming!"—a phrase widely heard in the media at that time, from the front lines in Vietnam.

Over the next few weeks I noticed that during my meditation sessions, if I concentrated awareness within different physical/spatial locations within my body, such as the heart or the throat, perceptually different sounds would arise in different locations and patterns, though the sounds were most clear and pronounced in the central region of my brain.

I soon concluded that the source of these perceived inner sounds must be of an electromagnetic nature, possibly the vibrations of a neuronal plexus within my nervous system resonating with electromagnetic modulations of our Earth's electromagnetic energy fields, or in the case of vending machines, the harmonic frequencies of some internal electrical radiation emanating from their circuitry, transformers, and so forth.

In bookstores I began to browse through books on anatomical structures of the brain and the central nervous system. This was the age before the internet, but luckily, I was living in New York City, and had access not only to the New York Public Library, but to many bookstores with medical sections. I was soon able to obtain excellent material with technical illustrations and X-ray photographs of internal physiological structures. I used these to visualize, with as much detail as possible, those internal areas, usually corresponding with the Indian chakra system, while meditating in the dark.

Over several years this process of concentrating and visualizing within areas of my body while focusing on the sound tones as they would arise became a main source of meditative practice for me, and the inner sounds tones grew ever more richly complex and often markedly louder in volume, and began to produce distinct tactile sensations of a flowing nature, unlike the sensations felt in the external senses of touch, vision, taste, and hearing.

On weekends I would also search for books for guidance in silent meditation, and in the process discovered Patañjali's *Yoga Sutras*. My first copy was a translation with commentaries by Professor Ernest E. Wood (1883–1965), having the rather impressive (and long) title of *Practical Yoga, Ancient and Modern, Being a New, Independent Translation of Patañjali's Yoga Aphorisms, Interpreted in the Light of Ancient and Modern Psychological Knowledge and Practical Experience*.[10] I was thoroughly impressed that Wood had first been educated in the "hard" sciences of chemistry, physics, and geology, and only later had he become so thoroughly fascinated by yoga and meditation that he undertook to become a Sanskrit scholar. Wood's translations of the *Sutras* seemed to me to be the perfect manual for the type of meditative exploration that had become my passion. After carefully studying Wood's translation for several months, I found a different translation of the *Yoga Sutras* by a professor holding a Ph.D. in chemistry, I. K. Taimni (1898–1978).[11] To my surprise, many of the translations and commentaries between the two books were markedly different. This led me to attempt an understanding of each word in the context of my own experiences and practices.

A Peyote Ceremony for Leonard Crow Dog

At that time, although I was becoming increasingly involved in studying and practicing yoga, the heightened focus of the press on the Oglala Sioux shaman, Leonard Crow Dog, intrigued me, and I was drawn to all I could find of Native American mysticism and culture. In 1973 I was invited by a good friend, the actress Johanna Adler, to join a small group of friends for a Saturday evening in a Native American Church* ceremony on the top floor of a building overlooking Washington Square. The impromptu ceremony was led by Leonard Crow Dog, the shaman for the Oglala Lakota who that year took over the village of Wounded Knee to demand that the federal government uphold treaty rights. We were told that most of the participants would be taking acid or peyote, which they were urged to take just before joining in the ceremony space. So, we found our way to a large sloped-floor classroom on the fifth floor of the vacant building that had recently been purchased by New York University. We had ingested little yellow Owsley acid pills ten minutes earlier.

The darkly lit room quickly filled up with about forty participants, and on a platform in the front, below, were two elderly men. Beyond them, a projector screen had been pulled down. The lights were dimmed further, and a projector came on as one of the men began to speak. A projector began filling the screen with images of early encounters between Native Americans and Europeans, at first showing peaceful encounters, the exchange of food and goods, and so on, but soon followed by scenes of violence against the natives. Many of the images seemed to be historical drawings and sketches. At some point the images became color photographs of contemporary Native American ceremonies involving bloodletting and chest piercings, as the drumming grew louder and faster. At that point we were fairly strongly under the influence of the LSD, and I recall becoming a bit paranoid, realizing I was one of a very few non-Native American people here. In the dark, surrounded by Native Americans, I began to feel a real sense of anxiety.

Suddenly the projector switched off, and we sat on the floor in the deep dark silence. After what seemed to be an interminable time of anxiety in the darkness, the drumming and singing of sacred songs began, with the familiar, "HAY-YA, hay-ay, hay-ay, hay-ay . . ." that I had heard in countless films involving Native American ceremonies. But time passed and nothing else happened. I wondered if my *wasichu* (white person) pres-

---

*The Native American Church (NAC), also known as Peyotism and Peyote Religion, is a Native American religion that teaches a combination of traditional beliefs and Christian symbolism, along with sacramental use of the entheogen peyote.

ence might be causing us to draw a blank, spirit-wise. Time crawled on and it seemed as if the drumming and chanting were growing louder, more intense.

And then it began . . . my awareness seemed to tune into a sort of psychic town hall meeting, as if I could hear the interacting thoughts of the shamans leading the ceremony. At one point I became certain that they were "discussing" me, or my soul or spirit, and debating whether I should be accepted into this confluence of minds. One distinct message seemed to be a query, something to the effect that, "Should we accept this one into our group, she is from a place up the Hudson River." And I got the unsettling feeling that they were not talking about me, but my mother's spirit, which clearly could be discerned as part of my own psychic holoflux. Indeed, my mother had been born and reared in the small town of Highland Falls, on the banks of the Hudson, not far from the West Point military academy. I sensed a distinct feeling-toned agreement of acceptance from this psychoidal group-mind of shamans, conveyed to me in sort of psychic shrug. Then the drumming resumed, taking us further into nontemporal, nonspatial dimensions of consciousness where the thoughts flowing were somehow "non-dual," there being no way to distinguish what might be my thoughts and feelings from those of the others participating in the group consciousness or shamanic psychoid.

After what seemed to be ages of time (or perhaps more accurately, dimensions devoid of time), we were abruptly catapulted through a new transformation of awareness as suddenly the drumming stopped, and a loud single note, coming from an electric guitar, filled the space! As the unrecognizable yet fluid soft-rock electric guitar and bass music welled up, some sort of light projector began to fill the darkness above with celestial images of stately moving galaxies! It was all so incredibly unexpected and unanticipated that it seemed our minds were just suspended in a void of galactic harmony, reinforced by the energy of the resonating amplified instruments.

As the soft early light of dawn began to drift in from the windows, the music subsided and eventually ceased, as the projected display of galaxies and stars on the ceiling and walls faded away. A tall man stood up in front of the shaman's stage to welcome everyone to share food outside in the early dawn of Washington Square. No one spoke as we gathered in the cool air of the quiet, New York City, Sunday dawn. Large chunks of fresh bread were passed around, along with bowls of nuts and seeds, and a large gourd of water.

Eventually Johanna and I got up and left for our own apartments. I recall walking along the dirty concrete sidewalks toward my apartment on the Lower East Side. We passed two transvestites walking slowly barefooted, holding their high-heeled shoes. Back in my fifth-floor walkup apartment, I tried to assimilate all that I'd experienced that night. One thing was apparent, that I had directly experienced how individual consciousnesses could merge into some larger entity of awareness, as if talking to itself. Clearly, I had a lot to think about.

Sanskrit Language and Patañjali's Yoga Sutras

The experiences of Little Sur Creek, the encounter at Hamilton Pool, and the Native American ceremony visions all reinforced my motivation to understand consciousness by seeking to experience and explore directly as many alternate regions of consciousness as possible. I became interested in all forms of meditation and began attending lectures and weekend workshops given by Indian and Tibetan teachers.

As a result of my experiences, I had discovered in the *Yoga Sutras* what seemed to me to be an exceedingly useful, highly detailed, and integrated set of instruction and theory for psychonautic exploration through the means of prayer, meditation, and specific psychophysical exercises. More than simply a detailed handbook of "how to meditate," Patañjali's *Yoga Sutras* presents a theory-practice continuum that includes, in the words of the scholar Ian Whicher, "effective definitions, explanations and descriptions of key concepts and terms relating to *theoria* and *praxis* in Yoga."[12]

Patañjali's sutras can be likened to a set of charts for navigation within the ocean of consciousness, much in the same way as the "rutters," written compilations of collected sailing experiences shared among Portuguese mariners, were used to cross oceans to new worlds prior to the development of scientifically calibrated nautical charts in the fourteenth and fifteenth centuries. The Portuguese rutters contained not only sketches, charts, and maps from firsthand accounts and direct observation; they also included a wealth of practical tips for exploring the New World oceans, pointing out such things as dangers to avoid, steering directions, and other practical instructions for those setting out upon ocean voyages. In a similar fashion, Patañjali's sutras can be viewed as a collection of practical information, a rutter for those choosing to leave their egos anchored behind and setting out for exploration upon the vast oceans of consciousness. Patañjali's *Yoga Sutras* is thus a compendium of integrated sailing instructions for the psychonaut. We are all immersed in various oceans of consciousness but, like fish swimming in water, to us the waters of consciousness are invisible. And like fish in water, we often experience confusion, both as individuals and collectively, when powerful, seemingly invisible currents of consciousness take hold and sweep us away to unfamiliar depths and strange regions of thought and sensation.

Realizing how important Patañjali's sutras were to understanding the dynamics of consciousness, I soon acquired additional English translations of his sutras, interpreted by more than a dozen individuals, each one expressing widely different interpretations of many areas of the *Yoga Sutras*. I soon realized that in order to develop my own understanding of this classic of contemplative yoga in the context of what I had been taught in physics, electronics, and communication theory, I would need to study Sanskrit and read the *Sutras* in the original.

After working my day job for five years as an electrical engineer in New York but becoming increasingly involved in exploring consciousness, with a primary focus upon traditional Asian meditation techniques, I made an abrupt change of course in my career by resigning my engineering job and moving to California to enroll in a graduate program of comparative Asian philosophy and religion at the California Institute of Asian Studies. Finally, I was able to study Sanskrit and Indian philosophy in a formal setting.

Ayahuasca and the Psychic Nanobots

After marrying and graduating from CIAS with my MA in Indian Philosophy, I began to focus on family life, raising my son and daughter. After several decades living with them in Saudi Arabia without using entheogens, decades in which my regular contemplative practice grew substantially in regularity and richer in experience, I returned to California and entered a doctoral program in San Francisco to study philosophy, cosmology, and consciousness studies. Having developed a network of like-minded colleagues, I was soon afforded the opportunity to participate in a small group working with a shaman from Ecuador to explore the use of the plant vine ayahuasca in a "healing meditation." The session was held at night with a dozen other participants in a large room in the middle of a redwood forest.

Beginning the ceremony around 9:00 in the evening, we ingested a "tea" made from the ayahuasca vine and sat in the dark evening silence waiting patiently to experience the effects of the plant medicine, as it was called. After about an hour and a half, I suddenly noticed that a large fan hanging from the ceiling above was writhing, and as I looked up I saw that what had been blades were now a huge, writhing black snake. At that point I had to urinate very intensely and got up to move toward the bathroom, but when I looked down I saw the floor was covered with a sea of smaller writhing black snakes, and I stepped on one and slipped backward. The next thing I recall is was waking up lying down on my back with the shaman fanning me with several large tobacco leaves and blowing tobacco over my face. Soon after I was able to stand up and make my way to the bathroom where I experienced the famous "purge" that often accompanies the consumption of ayahuasca, during which I felt an enormous amount of energy roaring up from my abdomen to my heart region.

After returning from the bathroom, I found a place to lie down and closed my eyes. While it seemed that several "stages" of awareness were being transited, I will relate only the most intense experience that I have been able to recall. At some point I felt that I was literally being taken apart. In fact, I was being so thoroughly analyzed, tested, tasted, and probed that I had a sudden fear that I might lose my "personality" or ego—that I would never be able to be the person I thought I was again. I felt as if I were being subjected to some cosmic plasma-magnetic entity rape, and for a brief time felt outrage, fear, and panic, but these feelings eventually subsided.

As the outrage passed, I began to sense that the "invasion" was of a benign character, having an almost caring quality. I was not being violated after all; instead, I was being "tuned" somehow at a very fundamental level, at what seemed to be a molecular, organic region or perhaps even lower, within a genetic or electronic region of my being. As I became increasingly calm, I experienced a sense of benevolent "tuning," and I began to realize why people speak of "plant medicine" in regard to ayahuasca. Although I have been exceptionally blessed with health and seldom have needed healing in the ordinary sense, I sensed that this "plant healing" or "tuning" had a different, deeper, psychophysical or neurophysical sense to it.

Visually what I was observing after reaching a state of tranquility was stunning! I was floating somewhere off to the side of and looking across (or down) at an enormous slab rising up from below and continuing upward, far beyond my perception, appearing like some behemoth intergalactic cruiser/city from a *Battlestar Galactica* or *Star Wars* set. It consisted of a myriad of complex linear/angular components that continued to dissolve and reform like hyperactive transforming robot component assemblies, changing as if modifying themselves through participatory resonance with my own interactive process of observing and being observed.

The lines consisted of sharp laser-like iridescent energy, thinner than a hair, and the flickering color of bright candle flames. I was floating alongside, viewing the phenomenon as a bird might view an enormous skyscraper flying fifty feet away from the building, the structure itself continuously forming and transforming with ultrathin iridescent bright lines that seemed antenna-like and angular, creating an enormous behemoth of a slab entity that seemed made of what can only be described as nanobot-like plasmodial shifting components.

They were definitely responding to my own presence, and indeed they seemed to be the source of the "probe" and "tuning" that was simultaneously going on so intimately within my own being, and which had initially made me panic. This probing went on for what felt like a very long time, during the latter part of which I found myself completely calm, receptive, and somehow grateful. In fact, I felt in some way that I was trading something with these plasma-magnetic entities or electrobiological nanobots; they were imparting to me a fine-tuning while receiving something in return.

For a long time it felt as if these spiritual nanobots were performing a kind of vehicle recall analysis or 100,000-mile checkup; tuning my many energy dimensions, physical and psychophysical systems; and prescribing and applying immediately the compensatory radiation-resonance tweaking treatments required in my resonating energy systems, perhaps at the quantum level, the DNA level, the molecular level, the cellular level, the neuronal level, the electromagnetic plasmoid level, and as yet unknown "levels" of being. Six hours later, however, to my surprise—and all of the next day and the following week—my feelings of well-being, alertness, and joy were all clearly enhanced, amplified, and seemed to be operating at much higher levels than normal.

This experience afforded me a new respect for the characterization of ayahuasca and other healing plants through their imparting to me a new understanding of the meaning of the concept of healing, which now implies a new urgency—a healing not only of me as an individual but of the entire ecology, the planet. I think something the vine energy-nanobot-plasma-magnetic entities did for me was strengthen what had been weak or broken links with the ecosphere. I feel an even greater concern for the planet and the animals and plants around me, and I feel a new enthusiasm for the study of consciousness in all its mysterious ramifications.

Part 2

THE PSYCHOPHYSICAL METAVERSE

4
Fechner's Psychophysics of the Human Soul

> *The consciousness with which the child awakes at birth is only a part of the eternal, pre-existing, universal, divine consciousness which has concentrated itself in the new soul. The physical universe and consciousness are co-eternal aspects of one self-same reality, much as concave and convex are aspects of one curve.*
>
> GUSTAV FECHNER (1801–1887)

William James, often referred to as the "father of American psychology," was deeply influenced by Fechner's work. James frequently cited Fechner's research in his own writing, well after Fechner's death, and in particular in his famous book, *The Principles of Psychology*. In 1904, James wrote an introduction to Fechner's book, *The Little Book of Life after Death*, enthusiastically summarizing Fechner's wide-ranging contributions as follows:

> Fechner's name lives in physics as that of one of the earliest and best determiners of electrical constants, also as that of the best systematic defender of the atomic theory. In psychology it is a commonplace to glorify him as the first user of experimental methods, and the first aimer at exactitude in facts. In cosmology he is known as the author

of a system of evolution which, while taking great account of physical details and mechanical conceptions, makes consciousness correlative to and coeval with the whole physical world. Fechner held the view that the entire material universe, instead of being dead, is inwardly alive and consciously animated. He believed that consciousness or inner experience never originated, or developed, out of the unconscious, but that it and the physical universe are co-eternal aspects of one self-same reality, much as concave and convex are aspects of one curve.[1]

Throughout recorded history one can find almost universal acceptance of the idea that each human being has a unique soul and that this soul is distinctly separate from the material body. Fechner agreed with this view and spent much of his life conducting research to investigate the premise using scientific methods. Yet among the generations of scientists following Fechner, there have been few interested in even discussing, let alone using science to investigate the concept of a pre-existing soul, primarily because of the conviction that human consciousness is a relatively recent phenomenon that has emerged from the evolution of matter. To the scientific materialist, it is the material world of space, time, and objects ("things") that is fundamental, not consciousness, and certainly not the soul. And yet it is ironic that the founder of the science of psychophysics, Gustav Fechner, so admired by James and a genius in mathematics and the physics of his time, included the human soul as one of his primary objects of study and speculation.

Fechner's Psychophysics

Fechner founded the science of psychophysics in 1860 with his publication of *Elements of Psychophysics* (*Elemente der Psychophysik*), a thousand-page discussion of results and conclusions from decades of experimental work in his laboratories in Leipzig to explore the relationship of physical stimuli to contents of consciousness. Fechner, born in 1801 in Prussia, originally received a scientific education, and yet he grew into a highly skilled psychologist, philosopher, musician, and poet. He received his medical degree at the age of twenty-one but turned immediately to the study of physics and mathematics. Fechner assumed that any future approaches to consciousness would include mathematical and physical underpinnings. In his book, he published the first mathematical equation to model human consciousness.

Yet, throughout his life, Fechner often went far beyond the fields of mathematics and physics in his published research, sharing his discoveries that give us a sense of his reported passion for understanding the universe, and in particular, consciousness, life, and death. He published the three-volume work, *Concerning Matters of Heaven and the World to Come*, in which, though trained as a scientist, he wrote openly concerning life after death.

Fechner believed in the continual dance of transformation of energy and consciousness leading to eventual reunion with the origin: "Only when death loosens the knot and removes the body which envelops every living soul, will there be added to the union of consciousness the consciousness of union."[2]

In his book *Nanna, or the Soul-Life of Plants*, Fechner argues that plants move, though we lack the patience to observe their movements. Preempting research into the chemicals plants emit, Fechner posits that plants communicate via a lexicon of fragrance. Like plant neurobiologists, he rejects the necessity of a nervous system for sensation and thought. That plants differ from humans and animals in structure and function does not prove they lack souls; rather, their souls assume a unique material form. In the same book, Fechner introduced the concepts of "Nightview" and "Dayview," a theme to which he returned often during the rest of his life. In his "Nightview," Fechner describes a materialistic and mechanistic view of the world. In contrast, his "Dayview" discusses an almost spiritual holistic perspective of nature, arguing that everything is interconnected and imbued with life and consciousness. With such a published view, Fechner is thought to have been among the first panpsychists; viewing the entire universe as being alive and replete with consciousness, instead of being dead matter as most of his colleagues assumed (as most scientists still do assume). And yet he was also a respected scientist of wide renowned. Through laboratory experiments, Fechner developed a surprisingly modern view of consciousness as waves of energy even before James Clerk Maxwell published his 1865 paper "A Dynamical Theory of the Electromagnetic Field," in which Maxwell predicted the existence of electromagnetic waves and proposed that light itself was an electromagnetic wave. Fechner also wrote about "souls," and believed that human beings stand midway between the souls of plants and the souls of stars, who are angels.

The focus of his work was to use the mathematics and the tools of science to discover the physics of the human psyche and sensory system that would facilitate the exploration of consciousness and assist in developing the means for communing with conscious entities, including human souls, in what he understood to be the normally hidden "spiritual" dimensions of reality. Fechner might be called a science-driven metaphysician, using the tools of physics to understand metaphysics. He felt it of the utmost importance to alter the direction that science had begun taking in his day, and he struggled to rescue science from what he saw as the completely unnecessary restrictions it had clearly begun to place upon itself by an increasingly exclusive, unquestioned adherence to the popular new (and engineeringly effective) mechanistic worldview.

In the history of psychology, the dynamic methodology that Fechner used, solidly grounded in the natural sciences, remains unrivaled, resulting in a number of remarkable and bold observations, many of which were discounted as pseudoscience by his contemporaries. For example, through his research and direct experience, Fechner came to believe that what are called "angels" are actually spherical beings of light that can communicate with humans through means of modulated light frequencies. He believed that the sun, moon, and planets all have their own unique sense of conscious "Self" into which conscious psychoids in their proximity eventually merge and reunite. Unfortunately, Fechner worked so extensively and consistently in order to directly experience metaphysical domains, that he also damaged his eyesight quite badly by staring for brief periods intensely at the sun, sensing that he was definitely receiving some sort of direct communication exchange in this way.

The Little Book of Life after Death

Perhaps Fechner's most fascinating metaphysical writings are to be found in *The Little Book of Life after Death*, which greatly attracted the interest of William James, so much so that James wrote an introduction to an English translation of the book when it was first published in Boston in 1905. James wrote:

God, for Fechner, is the totalized consciousness of the whole universe, of which the Earth's consciousness forms an element, just as in turn my human consciousness and yours form elements of the whole Earth's consciousness. As I apprehend Fechner (though I am not sure), the whole Universe—God therefore also—evolves in time: that is, God has a genuine history. Through us as its human organs of experience the earth enriches its inner life, until it also "get zu Grunde" and becomes immortal in the form of those still wider elements of inner experience which its history is even now weaving into the total cosmic life of God.[3]

What is most striking in Fechner's observations is his belief that the entire material universe is conscious in a wide range of wavelengths of energy. James agreed that "His belief that the whole material universe is conscious in diverse spans and wavelengths, inclusions and envelopments, seems assuredly destined to found a school that will grow more systematic and solidified as time goes on."[4]

In his writing he encourages metaphysical exploration in the same way that William James urged his colleagues to use "introspection" as a first-hand approach to gaining knowledge and information about human consciousness. Fechner instructed his students to silence the normal activities of one's thoughts, an admonition identical to the teachings of Vedanta, yoga, Zen, and other contemplative practices. With the thoughts constrained through practice, says Fechner, one can eventually begin to communicate directly with other conscious entities, both in spacetime and in the hidden dimensions, through the resonant communion of thought frequencies. Fechner says:

Stilled is all restlessness of thought, which no longer needs to seek in order to find. Rather begins now a higher interchange of spiritual life; as in our own minds, thoughts interchange together, so between advanced souls there is a fellowship, the all-embracing centre of which we call God, and the play of our normal thoughts is but tributary to this high communion.[5]

Yet unlike the metaphysics found in Buddhism or Vedanta, Fechner goes deeply into mapping a physics of consciousness using his knowledge of the material sciences, particularly in the understanding of consciousness in terms of wavelengths and frequencies, which imply the strong relationship of consciousness with electromagnetic fields.

Fechner speaks of a "psychophysical movement" in terms that are amazingly congruent with the "holoflux movement" described by Pribram and Bohm (see chapters 10 and 11). His ideas are also supported by Bohm's model that describes the wholeness of the universe as a selfsame interpenetration of the explicate order and the implicate order: "Consciousness or inner experience never originated, or developed, out of the unconscious, but it and the physical universe are co-eternal aspects of one self-same reality, much as concave and convex are aspects of one curve."[6]

For Fechner, the psychophysical movement has two major components, similar to Teilhard de Chardin's view (chapter 6) of a radial component and a tangential component to consciousness. Fechner denotes one movement as "psychical" (including desire, effort, love, etc.), and the other movement as "physical," in a domain using spatial terms that can be described mathematically in spacetime. William James writes about Fechner's approach to consciousness in terms of waves of energy:

> Speaking psychologically, we may say that a general wave of consciousness rises out of a subconscious background, and that certain portions of it catch the emphasis, as wavelets catch the light. The whole process is conscious, but the emphatic wave-tips of the consciousness are of such contracted span that they are momentarily insulated from the rest. They realize themselves apart, as a twig might realize itself, and forget the parent tree.[7]

Beyond the insulated individual thought, Fechner writes that there is a type of "higher thought" that is to be experienced by advanced souls. This higher thought is reminiscent of Teilhard de Chardin's noosphere, discussed in part 3 of this book. Fechner conjectures that individual human thoughts add into a "great spiritual organism" where "all individual ideas have their root in the universal, so at last will all souls, in fellowship with the highest, be absorbed into the divine." In fact, he says that the world of consciousness is like "a tree of souls, the root of which is planted on earth and whose summit reaches to the heavens."[8]

Though fascinated by the cosmological structure of consciousness in the universe, Fechner focused most of his life work on directly studying the human brain, human thought, and human consciousness. One of his most insightful observations was the likely existence of two personalities, two minds, in one brain.

Separate Minds in One Brain

One of Fechner's speculations about consciousness dealt with structure of the brain. He knew that the brain is bilaterally symmetrical and that there is a deep division between the two halves that are linked by a connecting band of fibers called the *corpus callosum*. Fechner speculated that if the corpus callosum were split, two separate streams of consciousness would result and mind, the individual, would essentially become two. Fechner assumed that his theory would never be tested. Yet in 1936, fifty years after Fechner's death, an American neurosurgeon completely severed the connections between the brains in each of three different patients, during the removal of tumors from the central corpus callosum regions. Since that time generations of split-brain patients have been studied postoperatively. Decades of information have given rise to a "dual consciousness" or "two minds" theory that explains, based upon the clinical evidence, that there must be two separate consciousnesses within the brain's left and right hemispheres. These two separate consciousnesses or perceiving mind centers often are seen to exhibit behaviors in competition with one another once the corpus callosum is severed.

In 2011 the Nobel Prize–winning psychologist Daniel Kahneman published a theory, based upon hard data, that is in essential agreement with one of Fechner's bold conclusions. Kahneman, in his Two Minds Theory,[9] observes that every human has two separate mental systems that operate independently of one another, a result that has often been observed in individual patients after the corpus callosum's 200 to 300 million hardwired axonal links have been surgically severed. The response of those with severed hemispheres reveals, in various social situations and tests, two very different personalities or responses, often in open, observable conflict with one another. One widely observed conflicting response has been called the "alien hand syndrome," frequently observed in patients with a completely severed corpus callosum. One hand can be seen to be reaching for an object to manipulate while the person denies, often verbally, that they do not want to carry out the action, even to the point of having to use the controllable hand to restrain the alien hand, as depicted in the film *Dr. Strangelove*.*

In his belief that splitting the corpus callosum would reveal the two separate "minds" in one brain system, Fechner's conjecture seems to have been validated within the twentieth century by multiple observations of split-brain patients, post-surgery, as well as the carefully researched theories of Kahneman and others.

Unfortunately, the detailed map that Fechner made of the metaphysical worldview was a very peculiar mixture of rather disparate traditions. His observations and conclusions, while brilliant and widely read, were seen to be idiosyncratic and exceptional, and yet, due to the increasing materialistic assumptions emerging within twentieth-century science, research in psychophysics became exclusively focused on the physiological mechanics of sensory systems, to the extent of excluding interest in sustaining the introspective, contemplative research required to explore consciousness directly that so intrigued Fechner (e.g., epiphanies, telepathy, dreams, hallucinations, communication with other "spirit" entities, the experience of pure consciousness beyond the senses, entheogenic-induced changes of consciousness, etc.). Fechner's essays display an intricate understanding of science-based pantheism, as can be seen in the following description of one of his conclusions.

> Fechner's pantheism explains what it means for a person, i.e., one's whole mind and soul, to be part of a superior subject. For this purpose, he developed the doctrine of the "psychophysical tiers of the world"; expounded in the greatest detail in *Psychophysics*, chapter 45. If we think of the oscillations of the physical bearer of psychical phenomena as exhibiting the form of a wave, whereby the psychical phenomena first enter awareness, after the strength of the waves rise above a certain level, namely the "threshold of consciousness," this

---

**Dr. Strangelove, or: How I Learned to Stop Worrying and Love the Bomb*, cowritten, produced, and directed by Stanley Kubrick, was released in 1964.

gives us a model for explaining how one consciousness is subordinate to another. God's consciousness is represented as a main wave. It carries smaller surface waves with shorter vibration periods and thresholds higher than that of the divine wave. There is discontinuity in the transition of one consciousness of surface wave to another, but there is continuity in God's main consciousness, whose threshold for awareness lies much deeper.[10]

William James was deeply influenced by Fechner's writings, which served to reinforce his own conclusions. In his book *The Varieties of Religious Experience*, he often cited Fechner, whom he felt to be a kindred soul in their efforts to approach the study of religion and mysticism with the tools of science and direct experience. James felt that the material sciences were losing valuable insights from refusing to include religion and mysticism in their purview, and he tried to stress the ontological reality of consciousness beyond the merely "understandable" world. As James writes:

> The further limits of our being plunge, it seems to me, into an altogether other dimension of existence from the sensible and merely "understandable" world. Name it the mystical region, or the supernatural region, whichever you choose. The unseen region in question is not merely ideal, for it produces effects in this world.[11]

James personally believed that each person has a soul outside of the material, a center that exists in what he called a spiritual universe, leading a person to behave as they do in the physical world. With the rise of behavioral science and growing skepticism regarding the use of direct experiential introspection as an appropriate tool for scientific research in psychology, the psychological establishment moved away from James's psychophysical theories, as they had from Fechner's theories of soul before James. Nevertheless, there is some hope that in the near future their approaches to soul and mind may again be taken up by a material science that seems to have reached a dead end when it comes to explaining the "hard problem of consciousness," as the Harvard-trained psychologist Charles R. Fox recently suggested:

Even though psychology began as the formal scientific investigation of the soul, such work was abandoned in favor of the mind. Contemporary psychology frequently ignores the mind in favor of behavior. Still, modern areas such as cognitive science, cognitive and systems neuroscience, and artificial intelligence are reviving the scientific interest in mind, and perhaps we will soon see Fechner's insight revived by a reintroduction of soul into modern scientific theory.[12]

A small number of twentieth-century scientists have indeed applied their skills, deep thinking, and often experiential research in the effort to understand and map the areas that so intrigued Fechner and James. Among these are Carl Jung (1875–1961), Pierre Teilhard de Chardin (1881–1955), and David Joseph Bohm (1917–1992); their contributions to psychophysics and the integrated map of consciousness that they have left us are described in detail in the following chapters.

5
Jung's Spectrum of the Psyche

Carl Gustav Jung's Topology of Consciousness

One of the most renowned and brilliant psychiatrists of the twentieth century, Carl Gustav Jung (1875–1961), who established the field of analytic psychiatry, made it his primary lifetime project to understand and map the phenomenon of consciousness and mind in the human being and the universe.

In his later writing, instead of beginning by focusing on the word *consciousness*, he chose to use the word *psyche*, an ancient Greek word that translates as "soul" or "butterfly" (fig. 5.1). In creating a model of consciousness, he began by redefining the word *psyche* in the context of his grand map of human consciousness: "By psyche, I mean the *totality* of all psychic processes, conscious as well as unconscious."[1]

Jung spent much of his long life working to develop an integrated model of the psyche and the dynamics of various regions of human consciousness operational in the cosmos. Born in Switzerland by Lake Constance, he was the oldest of two children and the son of a pastor. During his childhood he experienced a sense of being two personalities. In his diaries he called them "No. 1 and No. 2." He described No. 1 as being a gregarious thinker, a social child who loved reading novels, playing with other children, and studying science with enthusiasm, while he described No. 2 as preferring to be solitary, feeling a bit melancholy, and enjoying the experience of solitude and being in a state of communion with nature.[2] He felt that No. 2 was somehow more "real" than No. 1, and yet there was

Fig. 5.1.
The butterfly symbolizes for Jung the transformation of the human soul.

some conflict between his two personalities: No. 1 often wished to be free of the melancholy moods and the pervasive preference for isolation associated with No. 2.

During his university days, he continued to experience and critically observe the interplay of these two personalities. His ongoing self-introspection influenced his academic choices, and while he initially gravitated toward a degree in archaeology, he soon became fascinated with the new field of psychoanalysis as perhaps the key to understanding his experience of No. 1 and No. 2. He decided to pursue a medical degree, and in 1900 he began his study of medicine at the University of Basel. Jung was awarded his MD degree in 1902 and interned at a large psychiatric hospital where he began to acquire clinical data that helped him develop his model of the psyche.

The young clinician became fascinated by Sigmund Freud's psychoanalytic theories and applied for specialized training in psychoanalysis. In 1904 Jung sent Freud a copy of his recently published book, *Studies in Word Association*, in which he presented his research findings in the use of word association to identify the psychic connections of patients to contents residing in the unconscious regions of their psyches. This led to an exchange of letters between the two, and their correspondence eventually led, in March 1907, to their first meeting, shortly after Jung had relocated to Vienna. Freud soon regarded Jung as a talented young contributor to the new field of psychoanalysis. However, in 1908, at the age of thirty-three, Jung suddenly became aware of and fascinated by the field of mythology when he discovered how interpretation of seemingly mythological dream sequences afforded direct access to his patients' psyches. He began collecting and reading everything he could find regarding mythology in world literature. This new interest created a slight rift between Jung and his mentor, Freud. Around the same time, Jung gravitated toward a broad investigation of "spiritualism," which had become a lively topic in Vienna culture. In 1904 Rudolf Steiner had been elected leader of the Theosophical Society in Austria and Germany and worked to develop a synthesis between science and spirituality, a "spiritual science." Jung himself participated in numerous séances and began to think seriously about the structure of the psyche, based on his own firsthand experience of his No.1 and No.2, and Theosophical literature and lectures, which tended to be an amalgam of Hindu, Buddhist, and Western mysticism.

In 1911 Jung published *Symbols of Transformation*. The book delved into the symbolic aspects of alchemy and its relevance to radical psychological transformations he had observed in a number of his patients. However, his book fell outside of the interest of Freud, and seems to have accelerated the divergence of their interests. In spite of Freud's criticisms, Jung's reputation grew, and at the age of thirty-five he was elected to serve as the president of the International Psychoanalytical Association, a position he held for the next four years.

For the next twenty years, Jung continued to work as a psychiatrist and psychoanalyst. During this period, he established his own renowned analytical psychology practice in Zurich and gained recognition for his groundbreaking theories and therapeutic approaches. In 1933, he was awarded a full professorship at the Federal Polytechnic University in Zurich, where he held the chair of medical psychology until his retirement in 1941.

Over his entire lifetime he continued to develop his model of consciousness and the human psyche, and in 1946, at the age of seventy-one, he published a detailed version of his mature model in the essay "On the Nature of the Psyche." Jung's overall model of the psyche was groundbreaking owing to its expanded understanding of the unconscious (both individual and collective), its emphasis on individuation and wholeness, the introduction of archetypes and symbols, and the integration of spirituality and psychology. While his broad interdisciplinary approach struck many psychoanalysts as being "unscientific" owing to his inclusion of metaphysical and mythological "data," Jung's contributions significantly broadened the scope of psychology, and his insights continue to influence numerous fields of study to this day.

The following table offers a brief review of fourteen specific terms that Jung uses to elaborate his understanding of consciousness and the human psyche.

Ten Critical Components in Jung's Model of the Psyche

#	Component	Description
1	**The Psyche**	The **totality** of all psychic processes
2	**The Self**	The "sense of wholeness" and "identity" of the larger "Self" reflecting the entire Psyche; in contrast to the individual, separate "self" that reflects the seemingly separate individual Ego.
3	**Psychoids**	Basic elemental centers of consciousness, unrepresentable entities, building blocks of the greater psyche. From the Greek term *psyche*: the soul +*oid* from *eido*: shape or form.
4	**The Mind**	An information processing system through which psychic energies flow and sensory systems operate in spacetime; the *mind* include memory, logic, computation, and reflective awareness (analogous to current AI computing systems.)
5	**Instincts**	Rudimentary, low energy psychoids that provide programmed patterns (templates) to assist the mind in reacting to basic sensory stimulation and carry out reflexive actions in spacetime (e.g., breathing, heartbeat, flight or fight response, etc.)
6	**Archetypes**	Complex, high energy psychoids in which reside rich information-laden, feeling-toned patterns and templates of understanding that can act as sources of knowledge of an advanced nature.
7	**The Ego**	The integrated seemingly separate awareness; the personal consciousness complex that uses the mind as an instrument; the Ego reinforces the sense of being a separate "self," the sense of being a unique identity ("I").

Ten Critical Components in
Jung's Model of the Psyche (Cont'd)

#	Component	Description
8	**Personal Consciousness**	Psychoidally imprinted content that is accessible to the single individual Ego; one's personal history, experience, understanding, and acquired instincts.
9	**Collective Consciousness**	Psychoidally imprinted content that is accessible to all human Egos; collective (cultural) history, experience, and understanding.
10	**Complexes**	Elements of personal and collective consciousness that have been formed and reinforced over time both by individual experiences and relationships with collective psychoidal templates.

"Spirit" and "Life"

In his earlier writing, Jung often used the word *spirit* to indicate consciousness. For example, in 1926 Jung presented the paper "Spirit and Life" in which he observed that the human ego seems to be operational between the two terms frequently used in psychoanalytic discussions, and commented that "Life and spirit are two powers or necessities between which ego is placed."[3] He speculated that his own psyche consisted of a central ego, balanced between his No. 1 "life" dimension and his No. 2 "spirit" dimension.

In Jung's more mature model, the functioning ego in space and time draws its content from the much larger, normally hidden, regions of the psyche. Within the psyche exist the wider transcendent realms that are for the most part inaccessible to the ego. It must be understood that though Jung often refers to these vast bandwidths of the psyche as the "unconscious," it is only from the perspective of our ego's awareness that they are "unconscious." The collective psyche throughout all of its many dimensions is, indeed, fully conscious, but manifesting consciousness in a myriad of modes that are generally inacessible (and inconceivable) to the normally limited human ego.

Jung conceives of a visual map depicting the relationship (fig. 5.2). "Our present knowledge does not allow us to do much more than compare

Jung's Spectrum of the Psyche • 95

Fig. 5.2. Jung's intersection of the psychic regions with the material regions at the zero-point.

the relation of the *psychic* to the *material* world with two cones, whose apexes, meeting in a point without extension—a real zero-point—that touch and do not touch."[4]

The significance of this intersection point parallels the concept of the Omega point as defined by Teilhard de Chardin:, "Omega, in which all things converge, is reciprocally that from which all things radiate."[5] It is through each inconceivably small point of intersection that psyche projects out into the material world, while information from the material world reflects back through each point into the ocean of the psyche.

Here Jung situates the human ego at the intersection between psyche and material, between spirit and life (fig. 5.3), at "the still point of the turning world," as Eliot puts it. But these two regional domains, the psychic

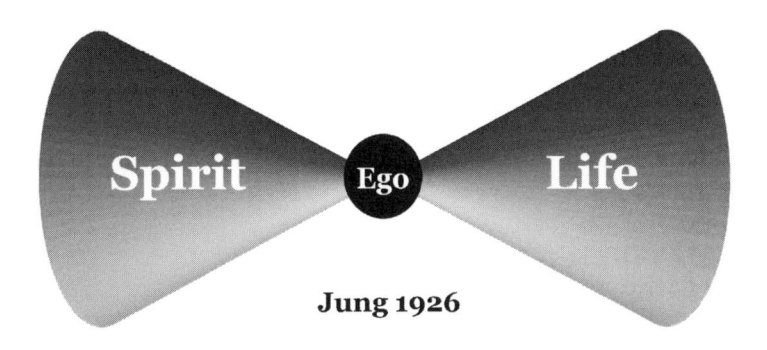

Fig. 5.3. Jung's early map of human consciousness: Spirit-Ego-Life.

(spirit) and the material (life), are not mutually exclusive. Jung sees them as being two different aspects on a continuum of the same single energy, or energy-consciousness as Vedantists would say. Jung sees them as one and the same thing:

> Since psyche and matter are contained in one and the same world, and moreover are in continuous contact with one another and ultimately rest on irrepresentable, transcendental factors, it is not only possible but fairly probable, even, that psyche and matter are two different aspects of one and the same thing.[6]

Much of the content within the dimensions of the psyche (spirit and life) Jung views as being "unconscious" to contrast it with what the conscious ego is able to perceive directly.

But Jung eventually became uncomfortable with the ambiguous popular use of the word *spirit*. After asking the question, "Is not the word 'spirit' a most perplexingly ambiguous term?"[7] Jung went on to list the variety of ways that psychoanalysts were treating the word *spirit* in their discussions:

- "Spirit" is often used to express an inexpressible, transcendental idea of all-embracing significance.
- In other instances, it is taken to be synonymous with "mind."
- Or it may be used to connote liveliness, wit, courage.
- Or it may refer to a ghost.
- Or it may refer to the dominant attitude of a particular social group.
- Finally, it can be used in a material sense, such as spirits of wine, ammonia, etc.

He then points out that in psychoanalysis the word should indicate precisely that aspect of consciousness that has a higher energy content and is somehow "above" life and the human ego.

> From the psychological point of view the phenomenon of spirit, like every autonomous complex, appears as an intention of the unconscious superior to, or at least on a par with, the intentions of the ego. If we are to do justice to the essence of the thing we call spirit, we should really

speak of a "higher" consciousness rather than of the unconscious, because the concept of spirit is such that we are bound to connect it with the idea of superiority over the ego-consciousness. The superiority of the spirit is not something attributed to it by conscious reflection, but clings to it as an essential quality, as is evident from the records of all ages.[8]

The interesting thing is that Jung supposes that only the ego is "conscious," and that life and spirit are realms that are unconscious to the ego. "Life" can be taken as the sum total of a myriad of processes (perhaps similar to modern computer software and physical mechanisms) that keep the many activities of the body operational but that are unconscious as far as ego is concerned (clearly, we are never really conscious of the physiological activities of our liver, spleen, hair growth, etc.). Likewise, the domain of "spirit" is usually unconscious, as far as the ego of the average individual is concerned (though perhaps not unconscious to the inner vision of mystics, contemplatives, and entheogenically fueled psychonauts).

The Psyche

In the sixteenth century, however, Renaissance writers began to substitute the word *psyche* for "soul," to reflect a more scientific description of the phenomenon. Similarly, in his mid-career, Jung also began using the word *psyche* in lieu of *spirit*, as can be seen in the titles of his major essays:

- "The Structure of the Psyche" (1931)
- "On Psychic Energy" (1948)
- "On the Nature of the Psyche" (1954)

The word *psyche* stems originally from Greek mythology, where Psyche was the name of a young woman whose name means "breath of life." Psyche was widely known as the "goddess of the soul." In the myth, Psyche falls in love with the god Eros (Cupid) and goes through a series of difficult trials placed in her way by the goddess Venus, the jealous mother of Cupid. Having overcome all but the final trial put in her way, Venus ultimately takes pity upon Psyche and transforms the human into a goddess replete with iridescent butterfly wings with which she is now able to join Cupid in his flights to the heavens, and eventually is able to marry Cupid. Thus the image of Psyche is often depicted in Greek art as having butterfly wings. The story can be found mirrored in the Christian idea of the human soul being tested (as in the story of Job) and eventually transformed through the acquisition of the love of self, neighbor, and God.

In his essay, "On the Nature of the Psyche," Jung elaborates his psychological model of the structure of the psyche. He expressed a growing interest in aligning his observations with the latest models from physics and mathematics, especially his view of what he had originally termed "synchronicity" that he had observed during a period in which he worked intensely with the *I Ching* oracle. Jung had first used the term *synchronicity* in 1930 to explain the Chinese approach to seemingly noncausal yet what seemed to be clearly interrelated, connected events.

The *I Ching* is a traditional Chinese divination system based on the principles of yin and yang, and the concepts of change and balance, light and dark, ascending and descending energies. The diviner typically asks a question, usually regarding something to occur or choices to be made that affect the future. To receive an answer from the higher dimensions of the universe, an interpretation of hexagrams is made, based upon what the final configuration of objects (three Chinese coins or fifty reedlike yarrow stalks) that are "thrown," tossed intentionally with the question in mind by the individual posing the question. The book itself, *The Book of Changes* (*I Ching*), is then used to assist in divining the outcome of the "throw." Jung became amazed at how often there seemed to be a direct valid connection between results of the *I Ching* throw and the unfolding of actual events in the real world.

Over the course of his therapeutic work, Jung encountered numerous instances where his patients shared stories of extraordinary coincidences and synchronistic events that held deep, even transformative, personal meaning. These accounts piqued Jung's curiosity even further and led him to investigate synchronicity as a possible physical-psychological phenomenon.

Jung and Pauli

Jung's interest in trying to understand the physical-psychological relationship of synchronistic events in terms of modern physics likely strengthened his close acquaintance and interaction with the Nobel Prize–winning pio-

Fig. 5.4. Carl Jung and Wolfgang Pauli, 1932.

neer of quantum physics, Wolfgang Pauli (1900–1958). In 1930 Pauli, a professor of theoretical physics in Vienna, had suddenly begun to experience the debilitating effects of a deep psychological crisis brought about by his wife's unexpected suicide, which was immediately followed by the death of his mother, and some months later, by the death of his favorite aunt.

Pauli first met with Jung in 1932, and soon after they entered into weekly sessions of deep psychoanalysis that continued with few interruptions over the next several years. After their therapeutical relationship seemed to resolve Pauli's problem, the two began to correspond with one another concerning the structure of the psyche, with emphasis in terms of what Pauli understood of quantum physics. During their years of psychoanalytic work, they discovered that both of them had on numerous occasions experienced synchronicity in events that occurred in their daily lives. Together over the next two decades of correspondence, they worked to flesh out a theory of synchronicity that they called the Acausal Connecting Principle. In 1952 the two co-authored a book on the subject, *The Interpretation of Nature and the Psyche,* and Jung's contribution was entitled *Synchronicity: An Acausal Connecting Principle.*[9]

One particular example of this phenomenon related to the recurring appearance of a specific number in Pauli's life, the number 137.

Throughout his life, Pauli had developed a close relationship with his aunt, and when she passed away, shortly after the death of his mother and the suicide of his wife, he fell into a deep state of depression. He related to Jung how for some inexplicable reason, the number 137 soon after began to appear in his daily life, repeated in various contexts.

The number 137 is also known to be a fundamental constant in quantum physics known as the "fine-structure constant." It is a prime number that has been derived from the speed of light, the charge on an electron, and Planck's constants. The number 137 is used widely by physicists to determine the probability that an electron will emit or absorb a photon. The frequent appearance of the number in unrelated events during this time struck Pauli as not only highly strange, but strangely significant in his life, and it became the theme of his psychoanalysis with Jung. The numerous synchronicities experienced by Pauli encouraged him to work openly with Jung so that together they were able to explore theoretical explanations regarding the potential connections between the physical world and the realm of the psyche. The two believed this experienced phenomenon would offer support to the philosophical idea of *nonduality*, the fundamental unity or oneness of existence, the essential interconnectedness and indivisibility of reality. In 1958, when Pauli died of pancreatic cancer, Jung was astounded to discover that the hospital room in which Pauli departed was numbered 137.

Location of the Human Ego

Jung had early in his career come to believe that life itself is primarily embedded in the material world, while, by contrast, the greater psyche somehow exists primarily beyond the limitations of the material world. He could only assume that the human ego had to be intersecting each these two domains: matter and psyche. Accordingly, he came to view the location of the human ego to be *at the intersection* of matter and psyche, with the ego functioning as a connecting bridge, linking the two regions (fig. 5.5).* Jung's material/psyche geometry is supported by the explicate/implicate geometry developed by David Bohm in his quantum physics that predicts an *explicate order* and an *implicate order*

---

*Note that earlier in his career this would have been equivalent to being located between "spirit" and "life" (see fig. 5.3, "Jung's early map of human consciousness: Spirit-Ego-Life").

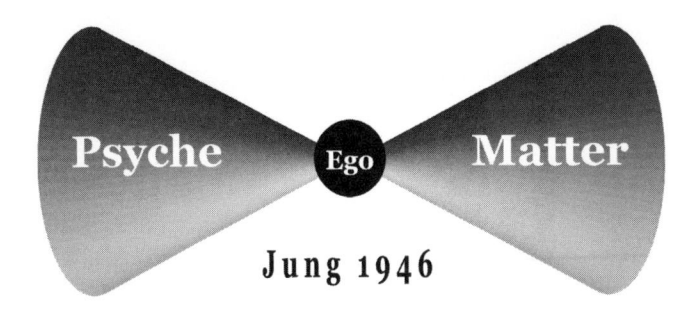

Fig. 5.5. Jung's new map of human consciousness: Psyche-Ego-Matter.

Jung became fascinated with trying to map the unconscious, the ocean of psychic content (psychoids) that the ego cannot perceive, but that can be contacted and observed during psychoanalysis. In trying to understand "the unconscious," Jung subdivided the *unconscious regions* of his model into five distinct "states of the unconscious."[10] These five states can be viewed as equivalent to "states of information" in the contemporary field of information processing (in the right-hand column of the table.)

Jung's Five States of the Unconscious

#Jung's "States of the Unconscious"	Human ego's "States of Information"	
1	Everything of which I know, but of which I am not at the moment thinking.	Information that is stored in the human psyche (known but currently out of mind; long-term and short-term memory).
2	Everything forgotten.	Information that is lost and irretrievable (forgotten, blocked).
3	Everything perceived by my senses, but not noted by my conscious mind.	Incoming sense information that is currently being filtered out of awareness (i.e., instinct-driven filters and preprogrammed censorship).
4	Everything that, involuntarily and without paying attention to it, I feel, think, remember, want, and do.	"Noise" (mental chatter: random thoughts, ideas, memories, churning through the individual ego-psyche).

Jung's Five States of the Unconscious (Cont.d)

#Jung's "States of the Unconscious"	Human ego's "States of Information"	
5	All the future things that are taking shape in me and will sometime come to consciousness.	Information still subliminal (undergoing development, growth, gaining energy, dimly apprehended, with potential to eventually emerge into awareness).

It is clear that all regions (with possible exception of no. 2) are dynamic and somewhat out of our conscious contact and control, although we have learned to modify them in various ways through a wide spectrum of conscious psychophysical practices and mind-altering substances.

Contents within Jung's "unconscious" regions of the psyche are not necessarily unconscious in and of themselves. Something is likely going on there that should be called consciousness, though of another quality or category as compared to normal human waking consciousness. However, these regions are not normally accessible to what is currently considered normal waking human consciousness, though during our sleep we can enter various regions (or channels) where what we call dreaming consciousness occurs. Indeed, much of Jung's contemporary psychoanalytic practice consisted in trying to bring contents from these "unconscious" regions into the spotlight of ego consciousness in an attempt to heal whatever fragmentation of consciousness has been causing pathological symptoms in his patient.

In a 1931 essay Jung describes his evolutionary view of consciousness within the psyche, pointing out that many of the various psychic functions found in today's conscious human psyche were once less conscious, in fact "were once unconscious and yet worked as if they were concious."[11]

What then is consciousness to Jung? From the previous quote, his tacit assumption is that it is *human consciousness* to which he refers, and that this human consciousness is not only variable and changeable but that it is in continuous evolutionary mutation, that it has arisen, grown, and expanded its regions or domains over human millennia, and that it is somehow tunable, that is, changeable or capable of self-programming of its own configuration.

The Psyche and Its Components: The Psychoids

As noted above, instead of beginning by focusing on the word *consciousness*, Jung chose to use the word *psyche*. In creating his model of consciousness, Jung began with redefining the word *psyche* in the context of the grand map of human consciousness. It is important to recall, as previously mentioned, Jung's inclusion of all modes of consciousness (of which there may be perhaps an infinite number of varieties). In discussing the psyche, Jung states, "I mean the *totality* of all psychoidal processes, conscious as well as unconscious."[12]

The building blocks of the psyche, Jung tells us, are what he terms *psychoidal processes* or *psychoids*, though the term is not widely appreciated by many psychoanalysts.

> Jung's psychoidal concept relates to a deeply unconscious realm that is neither solely physiological nor psychological. Seen as a somewhat mysterious and little understood element of Jung's work, this concept nonetheless holds a fundamental position in his overall understanding of the mind, since he saw the psychoidal unconscious as the foundation of archetypal experience.[13]

Jung conceived of psychoids as being the primary building blocks of all forms of consciousness, and in particular, he sees them as being integral components of the archetypes, those highly complex structural processes of consciousness that can directly affect the development and viability of human consciousness. Jung views psychoids as psychophysical energy patterns that act as building blocks for archetypes and that not only have active components existing in space and time, but simultaneously have components in, reach into, and span every dimension beyond those of space and time. For this reason Jung refers to them (and to the archetypes constructed by them) as "transcendent."

These psychoidal entities distributed throughout the universe as quasi-separate entities, subsets or projections of the one vast single consciousness he refers to as "the psyche." The manifest psyche can be seen as a multi-dimensional projection of quasi-independent psychoids that are also the building blocks of archetypes and all modalities of collective consciousness and the collective unconscious.

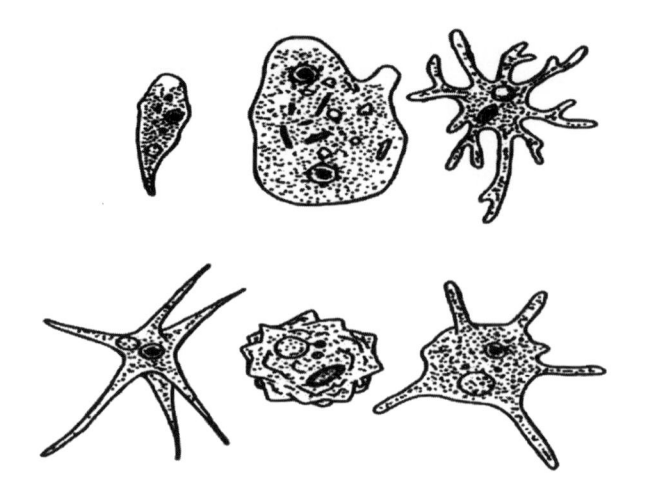

Fig. 5.6. Symbolic view of multidimensional psychoidal consciousness-entities.

According to Jung, psychoids are likely to be operational within every conceivable dimension of the larger metaverse, communicating with other psychoids through direct resonance. They are able to directly influence human behavior, development, and culture via linkages with individual human egos as well as collective consciousness of every sort.

Jung viewed the contents of the universal psyche as a multidimensioned projection of psychoids of innumerable types. Figure 5.6 depicts an artist's symbolic rendition of various multidimensional psychoidal entities contributing templates and information to the larger psyche. "The psychoid form underlying any archetypal image retains its character at all stages of development, though empirically it is capable of endless variations."[14]

In his early division of the psyche into unconscious and conscious contents, Jung's model of the psyche was somewhat binary. Jung explores variations of these binary configurations of the psyche, casting the psyche into several binary models including "Self" versus "Instinct" (now identifying the higher "Self" psychoid complex within the psyche) (fig. 5.7).

However in his 1946 essay it can be seen that his vision has evolved considerably, and his new understanding presents a more complex model of the psyche, one more closely aligned with the physics of electromagnetic fields. In this new model Jung considers the psyche to exist as a *spectrum of frequencies* (fig. 5.8).

By 1946, the radio had become widely adopted as a familiar means of communication. The growing public recognition that there exists a

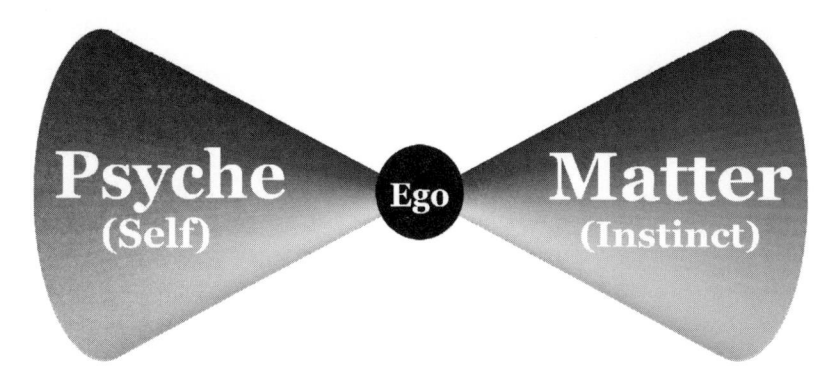

Fig. 5.7. Psyche (Self) –Ego –Matter (Instinct).

"spectrum of radio frequencies" that can be dialed (tuned into) on a radio offered Jung a new physical paradigm that was fully congruent with his understanding of the psyche existing within a wide frequency spectrum of possible manifestations.

Jung locates the psyche-spirit-Self at the higher ranges of the spectrum, higher frequencies at smaller spatial dimensions, while the biological instinctual psychoids are assumed to be at much lower frequencies (larger spatial dimensions).

Archetypes are psychoids exhibiting a dual nature, existing both in the implicate order (the psyche of an individual) and in the explicate order (the spacetime world at large). The archetype is not generally accessible to consciousness until two-way links have been established through directed awareness using symbols and liturgies.[15]

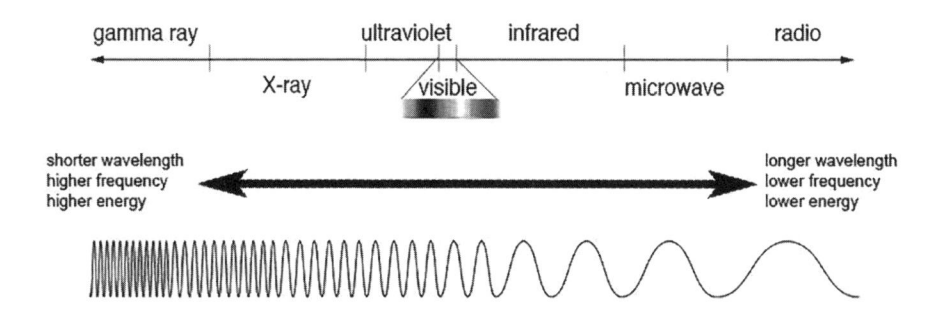

Fig. 5.8. Electromagnetic frequency spectrum.

The ego is able to intuit regions of the psyche such as personal memory and at times various otherwise unconscious contents through linkage to various psychoids.[16] Jung often refers to two regions of the unconscious, one that is "lower," consisting of instincts and rudimentary psychoids, and the other region containing more complex psychoidal entities of consciousness, which he calls archetypes (Jung tells us that many archetypes are referred to in religious texts as angels, demons, gods, etc.). Jung also came to believe that the psychic world (the domain consisting of multidimensional active psychoids) and the material world (the domain of space and time) are somehow distinct and separate, much as Bohm distinguishes the *explicate order* (spacetime) from the hidden dimensions of what he termed the *implicate order*.

The Light Spectrum: A Mathematical Metaphor for the Psyche

In his essays on the psyche, Jung recurrently expresses his longing for some "mathematical basis" to the psyche: "The tragic thing is that psychology has no self-consistent mathematics at its disposal."[17] However even without the mathematics he longed for, Jung proceeded to develop a mathematical metaphor for elements of the psyche along an axis analogous to the visible light region of the electromagnetic spectrum.

In his model he assigns instinctual psychoids and archetypical psychoids to distinct regions at opposite ends of the spectrum: "the biological instincts" are mapped to the lower frequency range of the spectrum (to the right in fig. 5.9), while he maps "the archetypes" to the ultraviolet region (to the left in fig. 5.9). Note that compared to the ultraviolet, the infrared is a region consisting of much longer wavelengths, lower frequencies, and hence lower energy, as compared to the ultraviolet. This corresponds with his contention that the archetypal psychoids, being much more complex than the instinctual psychoids, require higher frequencies and contain greater energy than do the less complex instincts. "Just as the 'psychic infrared,' the biological instinctual psyche, gradually passes over into the physiology of the organism and thus merges with its chemical and physical conditions, so the 'psychic ultraviolet,' the archetype, describes a field which manifests itself psychically."[18]

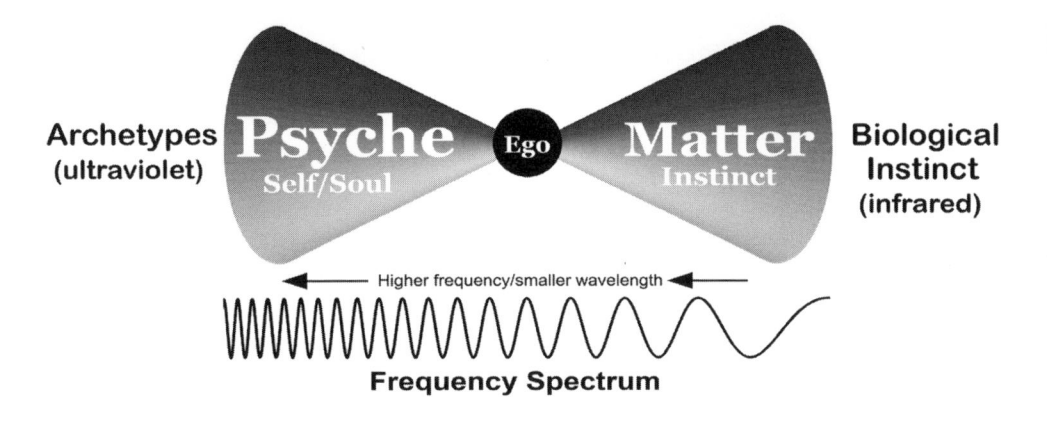

Fig. 5.9. The spectrum of consciousness: archetypes to instincts.

Based upon innumerable phenomena that he has himself observed during his years as a psychoanalyst, Jung supports his analogy of a frequency continuum (spectrum) of consciousness:

> So far as I can see at present, parapsychological phenomena are completely explicable on the assumption of a psychically relative spacetime continuum. The "psychic infra-red," the biological instinctual psyche, gradually passes over into the physiology of the organism and thus merges with its chemical and physical conditions.[19]

Jung not only relies on his image of a frequency spectrum as analogue to the psychic spectrum but frequently mentions the connection between energy and psychic processes in the physiological substrate:

> The psyche is not a chaos made up of random whims and accidents, but is an objective reality to which the investigator can gain access by the methods of natural science. There are indications that psychic processes stand in some sort of energy relation to the physiological substrate. In so far as they are objective events, they can hardly be interpreted as anything but energy processes.[20]

How a Sine Wave Is Created

Electromagnetic waves are depicted in the familiar sine wave graphical notation, the sinusoidal patterns commonly viewed on a two-dimensional graph (fig. 5.10). In actuality these waves are generated by the movement of electric charges in their motion circling a central location (a path similar to that of the planet Earth moving in a circular path around the sun). Imagine that the electron charge sends out radiant energy that is observable from a location far away. When the electron is closest to the observer (point A in the figure), the signal is strongest (brightest). When the electron is farthest away (point C in the figure) the signal is weakest (less intense) because it is farther away. A graph of the intensity of the observed signal creates a sine wave pattern as the electron orbits about

How a Sine Wave is Created

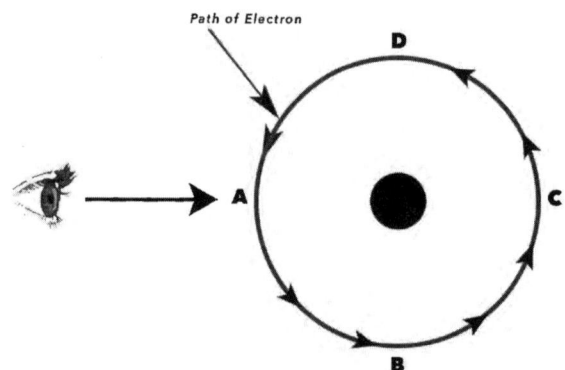

Point A is closest to the observer and the electron will be at maximum brightness
Point C is farthest away from the observer and the electron will be at minimum brightness Points D and B are midway between the two extremes.

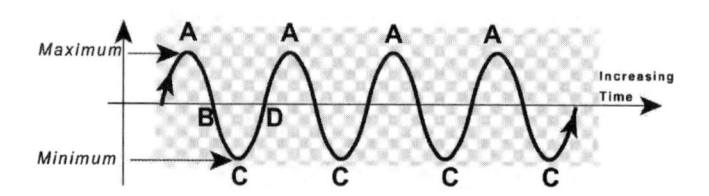

Fig. 5.10. How a sine wave is generated by rotation about a center.

its center of gravity (bottom of figure). The wavelength of the resulting electromagnetic signal is simply the diameter of its orbit, the circle in the figure.

The frequency of the electromagnetic signal is the number of revolutions the electron makes in a specified time (using the "human-time second," though a human second is actually 10^{44} "Planck-time-constant seconds"). Smaller orbits (smaller wavelengths) result in higher frequencies, because the electron, traveling at the speed of light in all cases, has a shorter path (circumference) to traverse.

Thus, electromagnetic waves are characterized by both *frequency* and *wavelength*. The wavelength corresponds to the diameter of the path (circle) about which the generating source charge spins. Thus, in electromagnetic fields, frequency and wavelength of its components are inversely proportional to one another (i.e., the shorter the wavelength, the higher the frequency, and vice versa).

A magnetic energy field is generated when an electric charge rotates around a central point (fig. 5.11). This is the principle by which commercial electric energy is generated. The magnetic field is perpendicular to the electron's path of rotation. Faster rotations (more rotations per unit of time), that is, higher frequencies (smaller wavelengths), result in a stronger magnetic field along the central shaft of rotation.

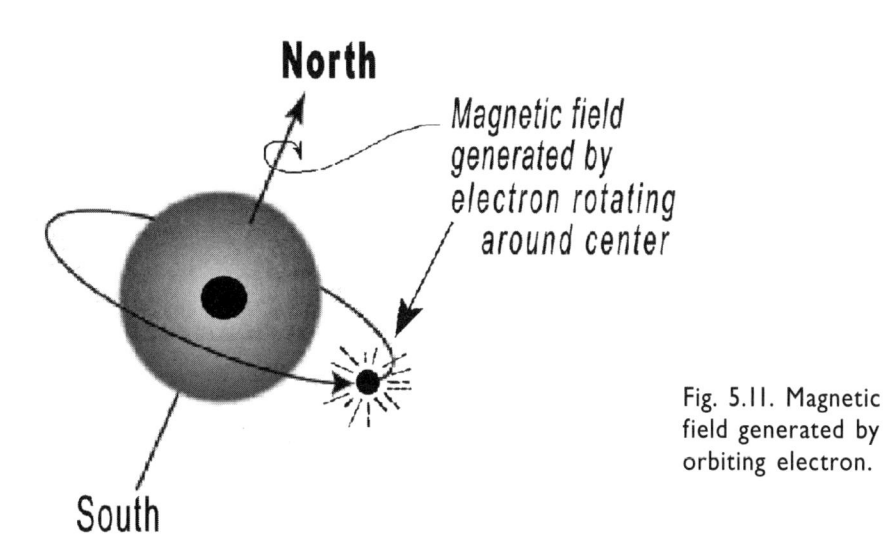

Fig. 5.11. Magnetic field generated by orbiting electron.

The extremely high frequencies that operate near the incredibly small Planck length generate stronger magnetic energy fields as compared to any longer wavelengths. Higher frequencies also offer higher information handling capabilities (think of the ever faster, smaller microchips that increase computing power).

Jung soon came to believe that *psychic processes must be* a phenomenon of electromagnetic energy and frequency:

> There are indications that psychic processes stand in some sort of energy relation to the physiological substrate. In so far as they are objective events, they can hardly be interpreted as anything but energy processes. The perceptual changes effected by the psyche cannot possibly be understood except as a phenomenon of energy.[21]

In our particular universe the range of sinusoidal wavelengths ranges from a maximum of the current diameter of the universe (approximately 10^{26} meters but increasing at the speed of light) to the smallest possible wavelength in spacetime at a diameter of the Planck-length limit of 10^{-35} meters, determined by the father of quantum physics, Max Planck in 1900. This presents an enormous frequency spectrum, though we humans are only able to view and/or manipulate an extremely tiny portion of the full spectrum of electromagnetic activity (fig. 5.12). Note that smaller dimensions (smaller wavelengths) are associated with higher energies of radiation.

Perhaps, as panpsychists believe, consciousness (Jung would say psychoids) of various sorts may exist at every wavelength (in every frequency range). In his discussion of the contemporary human psyche, Jung conjectures that the *complete* human psyche (i.e., the full range of consciousness from the instinctual unconscious and matter to the higher regions of Self), might be seen as actually spanning a specific portion of the spectrum, much as the visible light that humans can detect spans a specific region (though very small) of the entire spectrum of electromagnetic energy. He writes: "Psychic processes therefore behave like a scale along which consciousness 'slides.' At one moment it finds itself in the vicinity of *instinct*, and falls under its influence; at another, it slides along to the other end where *spirit* predominates."[22]

Jung's Spectrum of the Psyche • 111

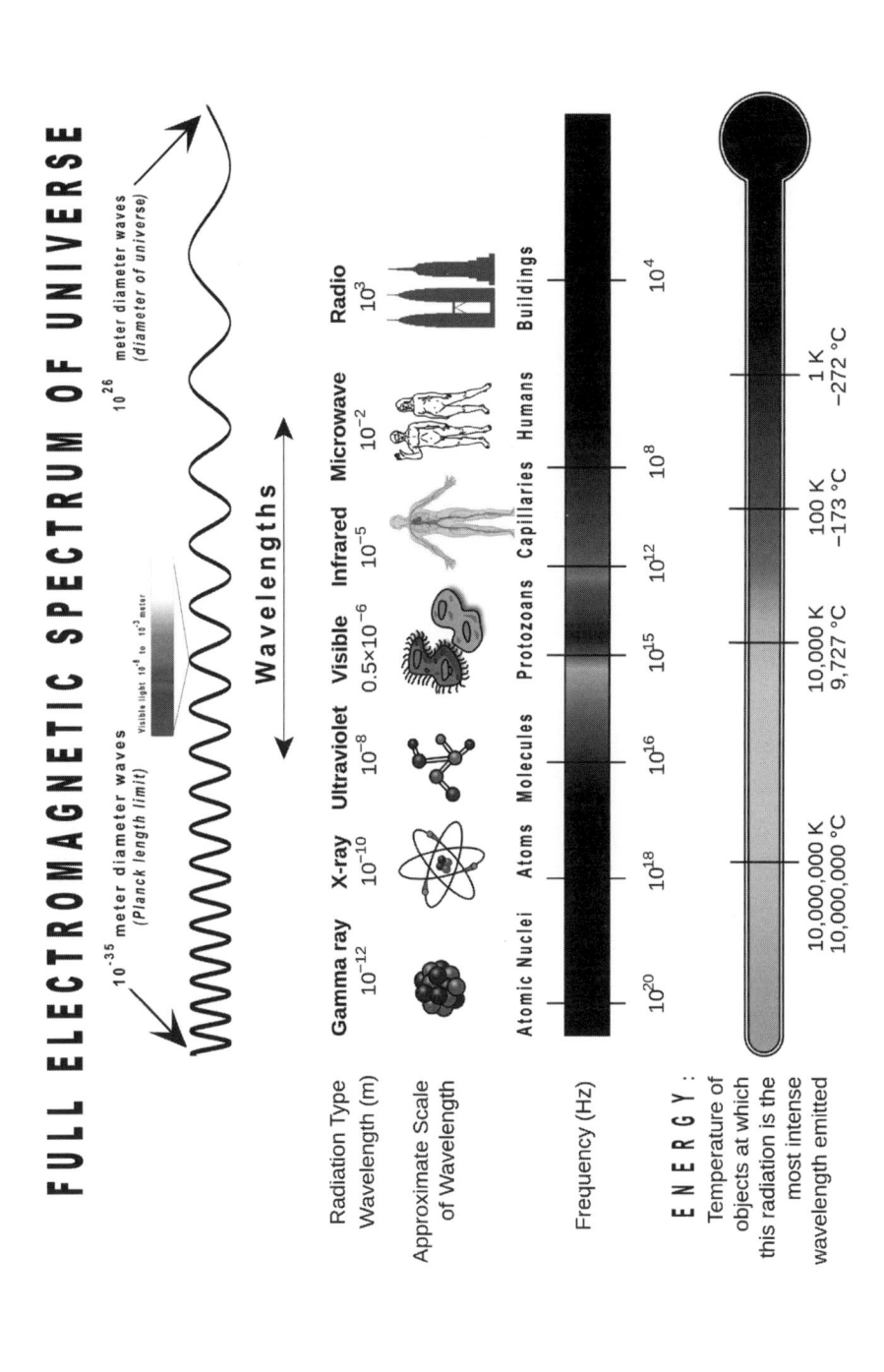

Fig. 5.12. Full electromagnetic spectrum of the universe (in wavelengths).

The Sun as a Psychoid

Jung conceived of psychoids as being psychophysical energy patterns of consciousness existing throughout the universe as quasi-separate entities, subsets or projections of the vast conscious psyche. These psychoidal entities are likely to be operational in every conceivable dimension, communicating with other psychoids of consciousness through direct resonance, and able to directly influence human behavior and culture through direct resonance with the individual human ego, effecting a consequent sharing of information once tuned links have been established between various psychoidal archetypes and psychoidal instincts.

The sun and its magnetosphere may be thought of as a psychoid, as it is composed of an enormously powerful magnetic field stretching out to the heliosphere, a vast bubble-like region of swirling magnetic energy that surrounds the sun and extends far beyond the orbit of Pluto. Recently several scientists have proposed that the sun may be conscious. In 2021 the biophysicist Rupert Sheldrake published an article, "Is the Sun Conscious?" in which he states that the sun may be able to sense what is going on throughout the solar system through the electromagnetic field that pervades the heliosphere, which could act as its primary sense-organ. Thus, the sun's mind could, in principle, know about all events within the solar system.[23]

Though much larger (in spatial dimensions) than a human psychoid, a sun psychoid would be one of innumerable quasi-independent psychic entities, all of which are subsets of their primary psyche. As our human psyche works through our mind (for memory storage and retrieval, logical manipulations, etc.), in a similar way, the psyche of the sun may work through its own much more energetic mind as Sheldrake surmises here:

> The mind of the sun, though centered in the sun itself, may integrate information from the entire heliosphere, just as our minds, centered in our brains, integrate information from our own bodies and the world around us. Similar principles may apply to countless other stars and solar systems.[24]

Archetypes

How then does Jung understand archetypes? While the ego is able to access personal memory and otherwise unconscious contents through linkage to various psychoids,[25] the elements of the higher, more complex archetypes that can manifest in spacetime are not generally accessible to a human ego consciousness until solid two-way links have been established through exposure of the ego to symbols and liturgical activities.[26]

The complex archetypal denizens of Jung's "higher" regions require higher information capacity than the more rudimentary psychoids that are found further down on the frequency spectrum energy chain. These higher archetypes manifest in space at increasingly smaller dimensions (recall that energy at smaller wavelengths operate at higher frequencies) than the "building block" psychoids. Complex archetypes (gods, demons, angels) are likely to operate at wavelength dimensions that may even be below the size of what material science call the fundamental particles (i.e., protons, electrons, muons, etc.). Psychoids "come in all sizes" relative to the wavelength of the energy bands in which they resonate. Paradoxically, the highest energy regions exist at the smallest dimensional wavelengths in space, conceivably even down at the bounding Planck limit to space where energies approach infinity.

Archetypes then are seen to function within the higher frequency electromagnetic spectrum in contrast to the more simple psychoids that reside within the lower frequency regions. For example, *instinct psychoids* (such as those that control breathing, heartbeat, and organs, etc.) likely operate as packets of wavelength in the radiant energy spectrum of longer wavelength packets found throughout human physiology. It is likely the vast human network of instinctual psychoids (various subroutines, from an IT point of view), that need to communicate and coordinate throughout the human physiology in order to maintain and operate the human body is in the range of far infrared, since that is the observed bandwidth of electromagnetic radiation (radio waves, if you will) an idea promptly dismissed by contemporary neurophysicists who see the phenomenon simply as human body "heat."

In figure 5.13 (p. 114), elements in Jung's map of the psyche can be viewed in a visual metaphor that includes all of the various psychoidal components of the psyche that he was able to identify during his sixty years of research and psychoanalytic experience with a wide range of patients.

The map depicts thoughts as a flock of birds flying high above the human

114 • The Psychophysical Metaverse

"Light of Day"
("Awareness")

Thoughts

Ego
Personal Consciousness
Social Collective Consciousness
- tribe, location
- family, ethnicity
- religion, liturgy
- politics, sports
- occupation, class
- social media

"Ocean of Consciousness"
(Higher frequency/energy collective,
"unconscious" to Ego's Personal Consciousness)

"Ocean of Consciousness"
(Lower frequency/energy collective,
"unconscious" to Ego's Personal Consciousness)

Archetypes   **Instincts**

Frequency Psychoids exist throughout the entire electromagnetic spectrum

Full spectrum of consciousness

10^{-35} m 10^{26} m

**Note: Smaller Wavelength means Higher Frequency

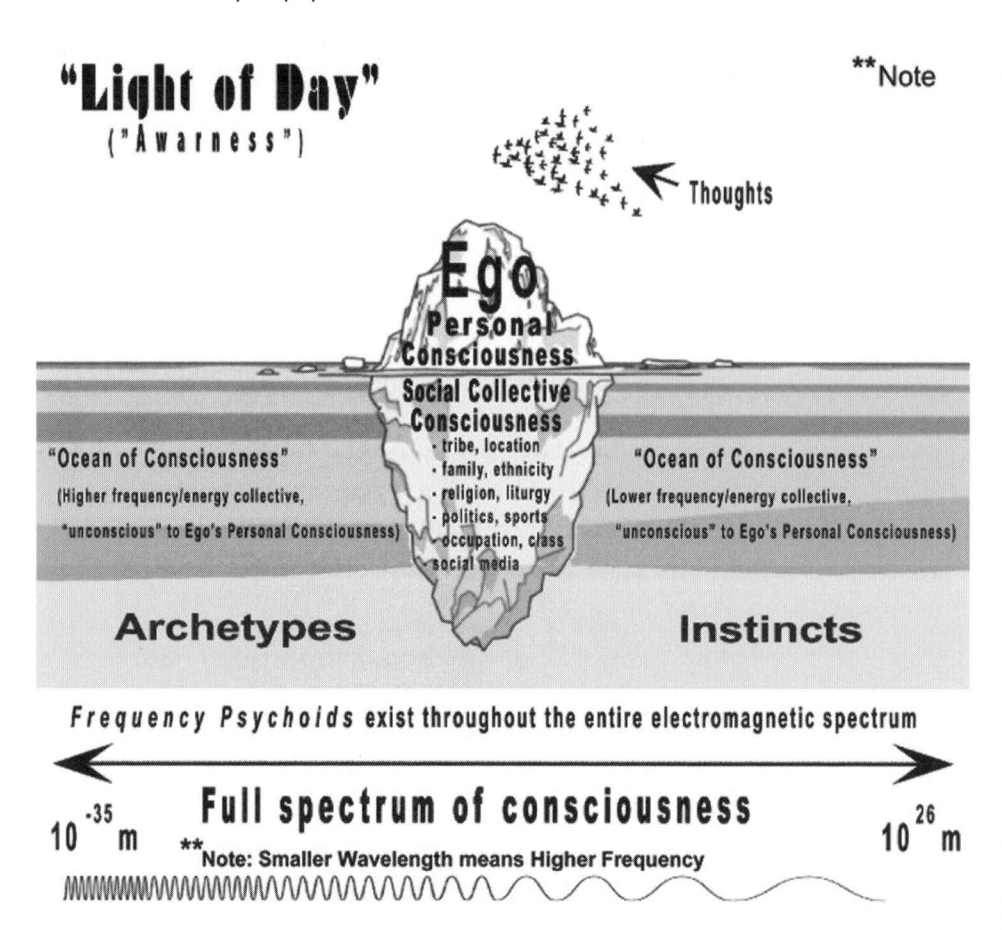

**Note

Fig. 5.13. Oceans of the psyche (a metaphorical map).

ego personal consciousness. This image comes from a lecture by Frederich Spiegelberg, one of the founders of CIIS, in which he describes how, when he is thinking or lecturing, he "reaches up mentally" and selects an idea from what seem to be a flock of birds somewhere high above his head.

To the left in the figure, the archetypes can be seen to be located at the high end of the frequency spectrum (shorter wavelengths) in agreement with Penrose and Hameroff's theory of "orchestrated objective reduction" (10^{-35} m). Instinct psychoids are shown at the right, in the region of lower frequencies, possibly in wavelengths in the region of neuron diameters (10^{-4} m). In Jung's essays, he chose to use the analogy of the human visible color range, from ultraviolet at the high frequency end of the psyche's spectrum, down to the infrared at the low end of the spectrum (lower frequency).

The *archetype* as such is a psychoid factor that belongs, as it were, to the invisible, ultraviolet end of the psychic spectrum. It seems to me probably that the real nature of the archetype is not capable of being made conscious, that it is transcendent, on which account I call it *psychoid*.[27]

This then is Jung's answer to the "hard problem" of consciousness as posed decades later by the Australian psychologist/philosopher David Chalmers:

A connection necessarily exists between the psyche to be explained and the objective spacetime continuum. The relative or partial identity of psyche and physical continuum is of the greatest importance theoretically, because it brings with it a tremendous simplification by bridging over the seeming incommensurability between the physical world and the psychic from the physical side by means of mathematical equations, and from the psychological side by means of empirically derived postulates—the archetypes—whose content, if any cannot be represented to the mind. Archetypes are typical forms of behavior which, once they become conscious [resonate within the ego], naturally present themselves *as ideas and images*.[28]

Archetypes as Energy Processes: Templates, Instructions

Jung sees the primary role of an archetype as being its *ability to organize* and a source of *guidance* and *knowledge*. According to Jung, an archetype is a psychoid that provides a pattern, a set of instructions, or the equivalent of software subroutines that inform and guide the form and formation of a human ego that has been able to connect with (resonate with) the archetype through the use of symbols, liturgical practices, or dreams. Archetypes guide existing processes that are evolving not only within the spacetime universe, but in the additional "hidden dimensions" as well. "The archetype itself (*nota bene*, not the archetypal representation!) is psychoid, i.e., transcendental and thus relatively beyond the categories of number, space, and time."[29]

Jung's vision is congruent with Sheldrake's understanding of morphic resonance. An archetype, through resonance, can affect entities (whether objects or psychic processes) by giving form to the growth and development to the object or process that has been able to establish resonance with the particular archetype. The archetype is basically an information transfer entity, a source of guidance and knowledge.

With regard to the frequency spectrum, Jung sees the higher psychoidal energy templates (called up through resonating consciousness through means of an image such as the Madonna, or Kali, or the Buddha) as having a locus of existence within the higher-frequency dimensions (bandwidths) of the unconscious. He metaphorically places the location of such archetypes in the violet end of the color spectrum.

Many different archetypes are able to resonate with the ego consciousness-spectrum of living beings to orchestrate morphic patterns of existence or process within space and time. During resonance between the archetype and human consciousness (whether individual or group consciousness), information content is transferred from the archetype to the human consciousness, after which (perhaps "coming out of a trance") the human consciousness (individual or group) once more continues as if completely separate from the collective archetypal consciousness. Jung here explains:

> This is completely explicable on the assumption of *a psychically relative spacetime continuum*. As soon as a psychic content crosses the threshold of consciousness, the synchronistic marginal phenomena disappear, time and space resume their accustomed sway, and consciousness is once more isolated in its subjectivity.[30]

To Jung, the ego finds itself in a bandwidth between the higher-energy-frequency collective unconscious and the lower-energy psychoids that offer instinctual templates to assist the ego in dealing with the myriad of daily challenges and the maintenance of vital life-sustaining functions. The "higher archetypes," when brought into contact with the ego, offer more complex patterns that are transferred to the ego. This transfer of informed energy empowers the ego with newly intuited knowledge that strengthens the ego's ability to solve the many complex challenges of behavioral existence as they arise in the spacetime world (and collective community) of the ego.

Yet a connection with an archetype can also be a double-edged sword and not always beneficial to the individual or collective group consciousness. Powerfully dark archetypes evoked by the Nazi swastika or a "great

Jung's Spectrum of the Psyche • 117

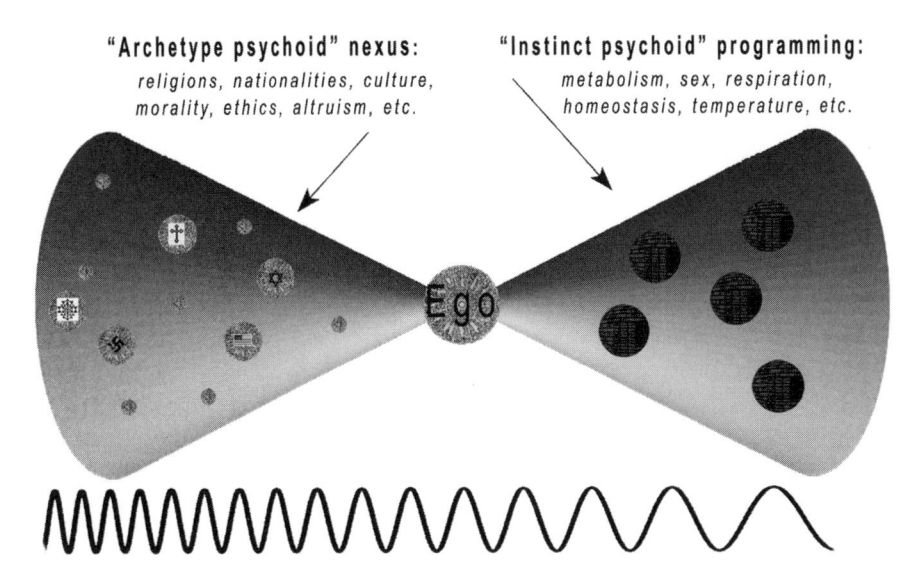

"Archetype psychoid" nexus:
religions, nationalities, culture, morality, ethics, altruism, etc.

"Instinct psychoid" programming:
metabolism, sex, respiration, homeostasis, temperature, etc.

Fig. 5.14. Archetype psychoids and instinct psychoids.

charismatic leader" come to mind. The metaverse appears to operate on ranges such as light/dark and good/bad, and individual (and group) consciousness entities can be strongly influenced by a wide range of archetypes, for good or bad. Furthermore, an archetypal influence can drown out and mask various contents of consciousness formerly operational in the ego that functioned to maintain balance and equanimity:

> The *archetypes* are formal factors responsible for the organization of unconscious psychic processes: they are "patterns of behaviour." At the same time they have a "specific charge" and develop numinous effects which express themselves as *affects*. The affect raises a particular content to a supernormal degree of luminosity, it does so by withdrawing so much energy from other possible contents of consciousness that they become darkened (by comparison) and eventually fall into the unconscious.[31]

The figure above (fig. 5.14) indicates the location of archetypes at higher frequencies to the left, while showing instinctual psychoids to the right. The central ego draws patterning from both regions (archetype and instinct) via frequency resonance.

It must be noted that Jung insists that the archetypes exist in dimensions that are normally inaccessible to the human ego band of the psyche.

Topologically, the ego can be thought of as an island rising above the surface of an enormous ocean of consciousness, vast regions of psychoids and archetypes, most of which are inaccessible to ego consciousness, and therefore, from the ego's viewpoint, unconscious. "We can hardly avoid the conclusion that between collective consciousness and collective unconscious there is an almost unbridgeable gulf over which the subject finds himself suspended."[32]

Jung believes that there is a *collective consciousness* that is identical in all individuals. This is quite in line with Teilhard de Chardin's concept of a *noosphere*, and offers support for Rupert Sheldrake's theory of morphic resonance.

However, Jung distinguishes between the broader collective unconscious and the personal unconscious. The archetypal collective unconscious is a vast ocean of psychic content that is available to all conscious entities, while the personal collective unconscious is made up of all the particular egocentric patterns and subroutines acquired by the evolving ego in its own particular temporal existence.

The following table indicates three collective ranges of the spectrum and articulates the influences upon each region. Jung describes how the various "collective consciousnesses" inform the ego:

Jung's 3 Ranges of Collective Consciousness

	The Psychecomponents of the Psychesources	
"Higher" Collective Unconscious	Archetypes "from angels to demons" (archetypal psychoids).	Planetary, solar, lunar, and galactic psychoids (gods, goddesses, angels, demons, spirits).
Collective Consciousness	Ancestral, tribal, social/ethnic collective (Teilhard de Chardin's noosphere).	Family and community, religious traditions, ethnicities, "isms," peer groups (social media constructs), social media hive consciousness.
"Lower" Collective Unconscious	Instincts; unconscious habits and triggered reactions (instinct psychoids).	Genetically transmitted subroutines (metabolism, procreation, respiration, sensory operating systems), reinforced habits, conditioning ("brainwashing"), trauma (PTSD).

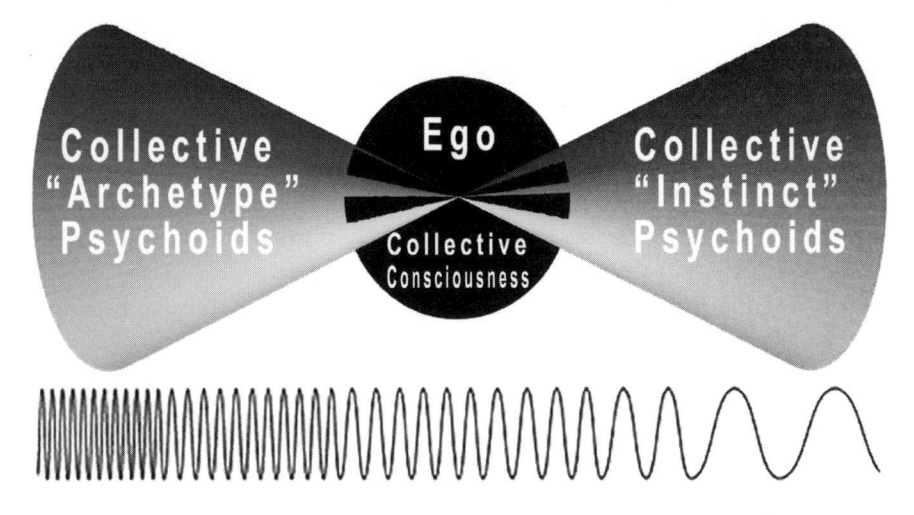

Fig. 5.15. The three regions of collective consciousness.

Ego-consciousness seems to be dependent on two factors: firstly, on the conditions of the collective, i.e., the social consciousness; and secondly, on the archetypes of the collective unconscious. The latter fall phenomenologically into two categories: instinctual and archetypal.[33]

Figure 5.15 shows the ego is in relation to three categories of collective consciousnesses.

The ego in the figure is depicted as having its own private collective consciousness. The ego is affected both by instinctual psychoids as well as whatever archetypal psychoids it has been able to contact and to resonate with in its own unique sphere of awareness. The archetypes affect each ego by imprinting patterns of knowledge, understanding, and power to the individual ego that has been able to "call up" (i.e., to make resonant contact with the archetype) through use of visually social (flags, personalities) and liturgical (crosses, stars, "emptiness") symbolism. Typically, these contacts allow the ego to transcend spacetime through the activation of such affective symbols and rituals.

Jung was fascinated by what he saw as the function of mythology and symbolism by which the ego could connect to, resonate with, and be influenced by the archetypes. However, in the modern world, the rise of secularism and the widespread loss of social engagement with religious and artistic symbolism has led to a loss of connection with the powerful archetypal influences that once formed and maintained individuals and societies in a healthy cohesive fabric. Traditionally effective healthy cultural symbolism has instead been replaced by relatively empty commercial "brand-name" imagery, personality cults, and nationalistic symbols having no direct connection or resonance with the numerous powerful healing archetypal forces that were contacted through traditional cultural activities and traditional religious practices.

Our contemporary global social collective consciousness, by having so weakened the link between human psyche and the numinous, means that billions have largely lost touch with the archetypes of the collective unconscious, the many "gods" of Greek and Roman tradition—as well as other polytheistic belief systems and mythology—that teach, heal, scold, and save humans in times of need. Without such links, the collective psyche has become primarily focused upon the material, upon wealth, and upon a myriad of sensory delights, rather than maintaining and developing healthy interactions with the transcendent dimensions that continue to exist everywhere and everywhen beyond space and time. Thus, the subjective consciousness of the ego has begun to gravitate away from independent connection with the greater healing (and healthy) archetypes and moved outward to join a wider collective public consciousness, in effect joining a larger "tribe" of consciousness that has its own rigid and unquestioned belief system. In fact, continues Jung, the more the ego identifies with the wider collective consciousness, the more it is cut off from the traditional archetypes.

> If the subjective consciousness prefers the idea and opinions of collective consciousness and identifies with them, then the contents of the collective unconscious are repressed. The repression has typical consequences: the energy-charge of the repressed contents adds itself to that of the repressing factor, whose effectiveness is increased accordingly. The more highly charged the collective consciousness becomes, the more the ego forfeits its practical importance.[34]

A strong collective religious tradition helped keep the medieval psyche in a relatively healthy balance; however, the contemporary loss of widespread religious views has resulted in egos that latch on to social "isms" as

Jung calls them. The function of religion, for the medieval human psyche, was that of *religere*, or a re-binding, re-suturing of the human ego-psyche to the archetypal energies that maintain balance, purpose, and creative transformative energy patterns.

> So long as the communal consciousness presided over by a Church was objectively present, the psyche continued to enjoy a certain equilibrium and constituted a sufficiently effective defense against inflation of the ego. But once Mother Church fell into abeyance, the individual became at the mercy of every passing collectivism and the attendant mass psychosis. One succumbs to social or national inflation. The tragedy is that one does so now with the same psychic mechanism which had once bound one to the church![35]

It can be said that "isms" flourish as *a sophisticated substitute* for the lost link with psychic realities formerly pointed to (bridged) by the archetypes through religion and public rituals (which today have largely been replaced by public sport spectacles and mass entertainment). The mass psyche that infallibly results *destroys* the meaning and worth of the individual and of traditional culture in general.

> If the ego proves too weak to offer the necessary resistance to the influx of unconscious contents and is thereupon assimilated by the unconscious, this produces a blurring or darkening of ego-consciousness and its identification with the preconscious wholeness, which amounts, therefore, to pathological effects. The psychic phenomena recently observable in Germany fall into this category.[36] [*Note*: this essay was written in 1931]

It is not only the individual ego that suffers from this failure to develop healthy connections with the higher archetypal patterns (psychoids), but the collective unconscious and the collective consciousness both begin to present pathological symptoms.

> From this it is clear that the isolated collective psyche not only disturbs the natural order but, if it loses its balance completely, actually begins to destroy its own creation. The present situation is so sinister that one

cannot suppress the suspicion that the Creator is planning another deluge that will finally exterminate the existing race of human beings [. . .] The archetypes are continuously present and active; as such they need no believing in, but only an *intuition* of their meaning and a certain sapient awe.[37]

Jung offers hope in his own self-observation of individuation, relating the experience of his own psychic being becoming whole: "Becoming whole has remarkable effects on ego-consciousness which are extremely difficult to describe. The ego cannot help discovering that the afflux of unconscious contents has vitalized the personality, enriched it and created a figure that somehow dwarfs the ego in scope and intensity."[38]

And so there is hope for the individual, even in today's world. Jung points out that some individual psyches are presently evolving toward even greater integration with certain powerful archetypes. Jung continues with a challenge:

Change of consciousness begins at home; it is an age-long process that depends entirely on how far the psyche's capacity for development extends. All we know at present is that there are single individuals who are capable of developing further, much beyond the common collective consciousness. How great their total number is we do not know, just as we do not know what the suggestive power of an extended consciousness may be, or what influence it may have upon the world at large [. . .] Is the time ripe for change, or not?[39]

6

Teilhard's Sense of Collective Consciousness

Teilhard's Integral Life Experience

In the first decades of the twentieth century, Marie Joseph Pierre Teilhard de Chardin,* a geologist, paleontologist, and priest by training, developed a model of consciousness that he referred to as "hyperphysics."[1] He conceived of this model in the context of his knowledge of physics, his keen observational skills as a geo-paleontologist, and his own specific introspective experience during a forty-year period of careful observation and consideration. In developing hyperphysics, Teilhard conducted an integral exploration of a region inclusive of and yet beyond the conceptual boundaries of physics or paleontology or the priesthood. The objectives of this chapter are to interpret and extend Teilhard's theories of hyperphysics in general, and to discover how Teilhard's theories can be reconciled with modern physics and contemporary theories of consciousness.

Henri de Lubac, eventually Cardinal de Lubac, was a Jesuit friend and correspondent of Teilhard's for more than thirty years. In a letter to Henri de Lubac dated 1934, Teilhard uses the word *hyperphysics*, describing it as a kind of metaphysics springing from the hard sciences, a metaphysics based upon science, yet "another sort of metaphysics which would really be a hyperphysics."[2] In the first sentence of the author's

---

*In this chapter, I refer to him as Teilhard de Chardin, or simply Teilhard.

note to *The Human Phenomenon* Teilhard states that the nature of his theories of hyperphysics is scientific, "purely and simply":

> If this book is to be properly understood it must be read not as a work on metaphysics, still less as a sort of theological essay, but purely and simply as a scientific treatise [. . .] only take a closer look at it, and you will see that this "hyperphysics" is not a metaphysics.[3]

It can be assumed that, trained in the sciences of geology and paleontology, Teilhard would have been an astute observer, constantly seeking and discerning patterns in the natural world. His published professional papers led to significant recognition in the field of paleontology; when published in 1971, a collection of his scientific papers filled eleven volumes.[4] Yet, in spite of the time constraints of his dual career as priest and scientist, he was able to develop, through a long series of unpublished essays written over his lifetime, a coherent theory describing a general physics of consciousness. Fueled by a lifelong practice of introspective observation, often alone in the silence of nature, Teilhard elaborated his map of the dynamics of consciousness—hyperphysics.

It is likely that Teilhard never knew how his life work in hyperphysics and the evolution of consciousness would be judged by mainstream science; the English edition of *The Phenomenon of Man* was only published in 1959, four years after his death, and was received somewhat critically both by the philosophical and the scientific community, as can be seen in this 1965 review: "Is his proposed 'hyperphysics' science? [. . .] There has been considerable confusion both in the United States and in Europe, where it appeared in French four years earlier, over what it is—physics, metaphysics, theology, mysticism, prophecy?"[5]

Rarely has a scientist, formally trained and active in a demanding technical profession, found time and interest (under the tacit threat of ridicule or censure) to develop a theory of consciousness based upon the data of firsthand participatory experience and observation. Pierre Teilhard de Chardin was one of the few: a trained scientist, holding a doctorate in geological paleontology from the Sorbonne, who wrote regularly and extensively to produce a science-based model of the evolutionary dynamics of consciousness, an "ultraphysics of union."[6] For over forty years, through

Fig. 6.1. Auvergne landscape. Photograph by Romary.

direct experience, observation, and keen analysis, Teilhard laid down a foundation and a legacy that he hoped would forge a new science-based understanding of the dynamics of consciousness in an evolving cosmos.[7]

Teilhard was born on May 1, 1881, the fourth of eleven children, in his family chateau, Sarcenat, in the most volcanic, mineral-rich region of France, Auvergne. Auvergne is also home to the largest oak forest in Europe,[8] and it is easy to see how the region of his birth instilled in Teilhard a passion for geology and nature (fig. 6.1).

Teilhard was a direct descendent of Voltaire, on his mother's side, and his father, descended from a noble family, was one of the largest landowners in the province, affording him ample free time to "build up sizeable collections of regional insects, birds, stones, and plants."[9] Teilhard soon became entranced with nature; he embraced his father's passion for geology, and at an early age began collecting and classifying specimens. But it was his mother's influence that instilled in her young son a love of the spiritual life, and he was introduced to regular family prayer, and a contemplative practice, the prayer on the Sacred Heart of Jesus. Following family tradition, Teilhard was sent to a Jesuit boarding school at the age of eleven.[10]

Of all the Catholic religious congregations, the Jesuit order is especially known for its emphasis on intellectual research and scholarship, and in this environment Teilhard excelled academically. But it was here, also, that Teilhard began his daily practice of sitting for an hour in silent contemplation:

It was at the school that he became an ascetic who voluntarily rose at dawn every day and went to sit in the chapel, often in freezing temperatures before the rest of the students awoke. He would follow a similar habit throughout his life, wherever he might be: in an Asian desert, in a prehistoric cave, or aboard a ship in rough seas.[11]

In 1899, at the age of seventeen, he made the decision to join the Jesuit order, and was formally accepted as a Jesuit novice, a candidate for eventual priesthood. After completing Jesuit secondary school, as part of his training as a novice, he was assigned abroad to teach physics and chemistry for a three-year period. Teilhard was sent to a Jesuit-run school in Cairo, Egypt, and it was there that he first experienced the fascination of an exotic new culture, an attraction to the mystery of surrounding antiquities, and most of all, perhaps, the experience of profound silence in the vastness of a desert.[12]

> It was immediately after I had experienced such sense of wonder in Egypt that there gradually grew in me, as a *presence* much more than as an abstract notion, the consciousness of a deep-running, ontological, total Current which embraced the whole Universe in which I moved; and this consciousness continued to grow until it filled the whole horizon of my inner being.[13]

Upon returning from Egypt, in 1909, Teilhard spent several years studying philosophy and theology at a Jesuit center in Hastings, on the coast of England, sixty miles southeast of London. It was during this time that he carefully read and reread *Creative Evolution*, a recently published work by the popular French philosopher Henri Bergson (1859–1941), a book the Vatican would soon place on its Index of Forbidden Books.[14] Previously, Teilhard had uncritically accepted the currently held theory of the *fixity* of species, and though he knew of Darwin's theories, they had seemed to him only an interesting hypothesis, one certainly suspect in the eyes of his Jesuit community. But after a careful reading of Bergson's *Creative Evolution*, Teilhard found himself suddenly a "convinced evolutionist," in strong agreement with Bergson's arguments for evolution, while yet disagreeing with Bergson's vision of a "pre-existing and obdurate matter" being operated upon by a life-force energy, which Bergson named *élan vital*. Teilhard himself felt that this life force, referred to by Bergson, was never to be found remote from matter, but from inception is at the very heart of matter.[15]

The dynamics of evolution, according to Bergson, is powered by a "vital" force of energy that animates not only life but the unfolding of the cosmos, and that fundamentally connects consciousness and body, an idea in radical contrast to widely accepted belief in the dualism of matter and consciousness, set forth by the seventeenth-century philosopher-scientist René Descartes.[16]

The young Teilhard was especially impressed with Bergson's emphasis on the importance of *intuition* and *immediate experience*, as these were Teilhard's own tools, developed and honed during his daily contemplative practices. Through immediate experience he was able to observe directly the structure and dynamics of his own inner space, his own complexity-consciousness in process, and he says that this led to "a new intuition that totally alters the physiognomy of the universe in which we move, in other words, in an awakening."[17] Near the end of his life Teilhard comments on the early influence of Bergson's book:

> I can remember very clearly the avidity with which, at that time, I read Bergson's *Creative Evolution* [. . .] I can now see quite clearly that the effect that brilliant book had upon me was to provide fuel at just the right moment [. . .] for a fire that was already consuming my heart and mind. And that fire had been kindled.[18]

The effect of Bergson's ideas upon Teilhard's worldview was significant indeed. In 1930, Teilhard wrote of Bergson, in a letter to his close friend Leontine Zanta (the first French woman to receive a doctorate in philosophy), "I pray for that admirable man and venerate him as a kind of saint."[19] After reading *Creative Evolution*, according to Teilhard's biographer Ursula King: "The magic word 'evolution' haunted his thoughts 'like a tune'; it was to him 'like an unsatisfied hunger, like a promise held out to me, like a summons to be answered.' Evolution was vital. It was the necessary condition of all further scientific thought."[20]

In 1912, Teilhard began formal graduate studies in geology and paleontology, eventually leading to his doctorate, though interrupted by

World War I; as a student, he also began working at the Museum of Natural History in Paris. In 1914 he was called up for service by the French Army and quickly trained as a medical orderly.[21] After serving for some time behind the lines, he volunteered to be reassigned to the Western Front, as a stretcher-bearer rather than as an army chaplain, and on January 22, 1915, he was assigned to a regiment of Moroccan light infantry, where "on arrival Teilhard made himself look like an Arab by exchanging his field-service blue for the khaki colors of the African troops, and his kepi for a red fez."[22]

It was here, alongside members of this regiment of Algerian tribesmen that Teilhard served for over three years in the trenches of the front lines. Teilhard was the only Christian in his regiment, but by the end of the war he was referred to affectionately by the North African Muslim soldiers he lived with in the trenches as "Sidi Marabout,"* an acknowledgment of his spiritual power as "a man closely bound to God, protected from all injuries by divine grace."[23] In 1921, at the request of his wartime regiment, Teilhard was awarded the French Legion of Honor for bravery.[24] The citation read:

"An outstanding stretcher-bearer, who during four years of active service was in every battle and engagement the regiment took part in, applying to remain in the ranks in order that he might be with the men whose dangers and hardships he constantly shared."[25]

Teilhard thus witnessed firsthand, for a protracted period of his life, the enormous suffering and destruction of human life that was the characteristic brutality of the war. Such experience was in sharp contrast to his academic life.

It was in Belgium that Teilhard experienced the true horror of World War I. When they arrived at Ypres, the troops found a town that had just been burned down. Hundreds of soldiers lay on the ground, dead or dying. And after the Germans were through with their conventional weapons strike, they attacked their enemy with poison gas.[26]

Yet the young scholar/priest seemed to display no fear, at least to his closest colleagues. One of his fellow soldiers at the time, Max Bégoüen, wrote the following, describing an event he witnessed on the Belgian front, in 1915:

---

*An Arabic title of great esteem and honor; *sidi* refers to a North African settled in France; *marabout* designates a saint and ascetic blessed with divine favor.

The North African sharpshooters of his regiment thought he was protected by his *baraka* (an Arabic word meaning "spiritual stature" or "supernatural quality"). The curtain of machine gun fire and the hail of bombardments both seemed to pass him by. During the attacks of September 2 at Artois, my brother was wounded, and, as he wandered on the battlefield, he saw a single stretcher bearer rising up in front of him, and he, for it was Teilhard, accomplished his mission quite imperturbably under terrible fire . . . "I thought I had seen the appearance of a messenger from God."

I once asked Father Teilhard, "What do you do to keep this sense of calm during battle? It looks as if you do not see the danger and that fear does not touch you."

He answered, with that serious but friendly smile which gave such a human warmth to his words, "If I am killed, I shall just change my state, that's all."[27]

For four years he served as an unarmed stretcher-bearer at Verdun, until 1917, and thereafter in the front trenches of Chateau Thierry in 1918, participating in action of such great ferocity that it took the lives of over nine million of his fellow soldiers. The accounts of his disregard of his own safety in order to rescue the wounded of all nationalities led to being his awarded for bravery in action.

We can only imagine the experiences this young man must have lived through, the sounds, the sights, the deprivations of weather and humanity, yet out of the intensity of this existential life "on the Front," Teilhard began to experience a new form of consciousness not only in his own but collectively, a "quasi-collective" participatory functioning flux becoming "fully conscious" as he describes in a letter written at the Western Front of the French lines in 1917:

> I'm still in the same quiet billets. Our future continues to be pretty vague, both as to when and what it will be. What the future imposes on our present existence is not exactly a feeling of depression; it's rather a sort of seriousness, of detachment, of a broadening, too, of outlook. But it leads also to a sort of higher joy, I'd call it "Nostalgia for the Front." The reasons, I believe, come down to this; the front cannot

but attract us because it is, in one way, the extreme boundary between what one is already aware of, and what is still in process of formation. Not only does one see there things that you experience nowhere else, but *one also sees emerge from within one an underlying stream of clarity, energy, and freedom that is to be found hardly anywhere else in ordinary life—and the new form that the soul then takes on is that of the individual living the quasi-collective life of all* men, fulfilling a function far higher than that of the individual, and becoming fully conscious of this new state [. . .] This exaltation is accompanied by a certain pain. Nevertheless it is indeed an exaltation. And that's why one likes the front in spite of everything, and misses it.[28]

During this time and under these conditions he began to experience, observe, and ultimately write about, his own direct awareness of nonordinary states of consciousness. For example, it was at the front that he began to perceive, directly, for the first time, a sense of collective consciousness over and above his own. In his 1917 essay "Nostalgia for the Front," Teilhard asks, "Is it not ridiculous to be so drawn into the magnetic field of the war [. . .] more than ever the front casts its spell over me [. . .] What is it, then, that I myself have seen at the front?"[29] And he answers himself, "it is above all something more, something more subtle and more substantial, I might define it as a superhuman state to which the soul is borne." Having left the front lines, he experiences a feeling of loss: "I have the feeling of having lost a soul, a soul greater than my own, which lives in the trenches and which I have left behind."[30]

It trying to understand the impact of these experiences on the young Teilhard, it is worthwhile to consider that at Verdun, where Teilhard served during one of the most protracted battles of the war, the single battle continued for over nine months, and the human losses approached apocalyptic proportion: "A French estimate that is probably not excessive places the total French and German losses on the Verdun battlefield at 420,000 dead, and 800,000 gassed or wounded; nearly a million and a quarter in all."[31]

It was at Verdun, the night before the attack on Fort Douaumont, on October 14, 1916, that Teilhard experienced an extraordinary vision, which he recounted afterward in the short essay, "Christ in Matter."[32]

Here, in the visual imagery alone, one can only imagine that Teilhard was experiencing a major psychotropic vision, perhaps brought on by fatigue, synesthesia from the constant bombardment, or from the stress of being continually at the front on the eve of a major offensive attack.

The Powerful Vision

The imagery of Teilhard's vision is so intense and specific here that one wonders, even, if he might have ingested ergot-infected rye bread (ergot mold on rye bread has been reported to induce LSD-like symptoms; in 1951 an entire French village became infected with the ergot alkaloid, experiencing hallucinations).[33] Here is Teilhard's description of this pivotal, altered-state experience that occurred in an abandoned chapel, at night, during the battle of Verdun:

> Suppose, I thought, that Christ should deign to appear here, in the flesh, before my very eyes—what would he look like? Most important of all, in what way would he fit himself into Matter and so be sensibly apprehended? [. . .] Meanwhile, my eyes had unconsciously come to rest on a picture that represented Christ with his Heart offered to men. This picture was hanging in front of me, on the wall of a church into which I had gone to pray [. . .] I was still looking at the picture when the vision began. (Indeed, I cannot be certain exactly when it began, because it had already reached a certain pitch of intensity when I became aware of it.) All I know is that as I let my eyes roam over the outlines of the picture, I suddenly realized that they *were melting*. They were melting, but in a very special way that I find it difficult to describe.
>
> If I relaxed my visual concentration, the whole of Christ's outline, the folds of his robe, the bloom of his skin, merged (though without disappearing) into all the rest [. . .] the edge which divided Christ from the surrounding World was changing into a layer of vibration in which all distinct delimitation was lost [. . .] I noticed that the vibrant atmosphere which formed a halo around Christ was not confined to a narrow strip encircling him, but radiated into Infinity. From time to time what seemed to be trails of phosphorescence streamed across it, in

which could be seen a continuous pulsing surge which reached out to the furthest spheres of matter—forming a sort of crimson ganglion, or nervous network, running across every substance. *The whole Universe was vibrating* [. . .] It was thus that the light and the colours of all the beauties we know shone, with an inexpressible iridescence [. . .] these countless modifications followed one another in succession, were transformed, melted into one another in a harmony that was utterly satisfying to me [. . .] I was completely at a loss. *I found it impossible to decipher* [. . .] All I know is that, since that occasion, I believe I have seen a hint of it once, and that was in the eyes of a dying soldier.[34]

Decades later, Teilhard refers to this epiphanic experience, this "particular interior event" of forty years prior.[35] He describes how it has been that, ever since this early revelation, he has had "the capacity to see two fundamental psychic movements or currents," which, when he first perceived them in his 1916 epiphany, "reacted endlessly upon one another in a flash of extraordinary brilliance, releasing [. . .] a light so intense that it transfigured for me the very depths of the World."[36] In his final essay, completed a month before his death, Teilhard stresses the *objective validity* of this initial evidence that had led directly to his new understanding of consciousness and the universe, evidence that had presented itself to him experientially in 1916:

> What follows is not a mere speculative dissertation in which the main lines of some long-matured and cleverly constructed system are set out. It constitutes the evidence brought to bear, with complete objectivity, *upon a particular interior event, upon a particular personal experience* [. . .] Today, after forty years of continuous thought, it is still exactly the same fundamental vision that I feel I must present, and enable others to share in its matured form—for the last time.[37]

It is this subsequent "forty years of continuous thought" that makes the uniqueness of his observations, expressed in his essays, so significant for the development of consciousness studies. Four sides of Teilhard's nature reinforced one another, integrally it would seem: scientific training, mystical vision, exceptional intelligence, and a passionate enthusiasm for discovery and understanding. While he was formally a scientist, highly trained and experienced in the observation, collection, classification, and written interpretation of geological and anthropological data, yet he was also a Jesuit priest, deeply immersed in observing the internal phenomena of spirit during his daily contemplative period. His was a quest to bring scientific reasoning and understanding to bear upon a direct vision, one that has been described by his biographer as: "A powerful vision linked to experiences of a deeply mystical, or what might be called pan-entheistic, character although he often simply called them 'pantheistic.' These experiences occurred over many years."[38]

Throughout his writing, one encounters passages that can only be seen to refer directly to personal experience of a sort of perception that he himself categorized as pantheism and mystical vision (which, along with his fascination with evolution, caused him enduring conflict with more conservative forces in the Vatican). Teilhard states that his perception, "as experience shows, is indeed the result [. . .] of a mystic absorbed in divine contemplation."[39] Elsewhere Teilhard regards this special psychic perception as a natural ability, but one that requires practice and cultivation in order to catalyze the required change of state in consciousness:

> This perception of a natural psychic unity higher than our "souls" requires, as I know from experience, a special quality and training in the observer . . . once we manage to effect this change of viewpoint then the earth, our little human earth, is draped in a splendor. Floating above the biosphere, whose layers no doubt gradually merge into it, the world of thought, the noosphere, begins to let its crown shine. The noosphere![40]

Teilhard's contemplative, mystical interests began at an early age, during which he searched to discern some "Absolute" in his experience of prayer with his large Catholic family, described here in, "My Universe," written on the battlefield of the Marne, three weeks after the beginning of a major attack by the Germans:

> However far back I go into my memories (even before the age of ten) I can distinguish in myself the presence of a strictly dominating passion:

the passion for the Absolute. At that age, of course, I did not so describe the urgent concern I felt; but today I can put a name to it without any possible hesitation. Ever since my childhood, the need to lay hold of "some Absolute" in everything was the axis of my inner life.[41]

During 1926 and 1927 Teilhard wrote *The Divine Milieu* while working in China, where he had effectively been banished by the Jesuit authorities; it is in the middle of this essay that he describes what can be only understood as a personal experience of deep contemplation in which, through a process of increasing centro-complexity, he began to travel consciously toward an encounter with a heretofore unimagined depth of inner being:

> And so, for the first time in my life perhaps (although I am supposed to meditate every day!), I took the lamp and, leaving the zone of everyday occupations and relationships where everything seems clear, I went down into my inmost self, to the deep abyss whence I feel dimly that my power of action emanates. But as I moved further and further away from the conventional certainties by which social life is superficially illuminated, I became aware that I was losing contact with myself. At each step of the descent a new person was disclosed within me of whose name I was no longer sure, and who no longer obeyed me. And when I had to stop my exploration because the path faded from beneath my steps, I found a bottomless abyss at my feet, and out of it came—arising I know not from where—the current which I dare to call *my* life. What science will ever be able to reveal to man the origin, nature and character of that conscious power? [. . .] Stirred by my discovery, I then wanted to return to the light of day and forget the disturbing enigma in the comfortable surroundings of familiar things.[42]

In the development of Teilhard's mystical sense, the possibility cannot be ruled out that Teilhard in midlife had the occasion to experience consciousness-expanding drugs, which would have provided new material for development of his theories of consciousness. On an ocean passage from France to China in 1926, Teilhard had befriended a French couple with a homestead in East Africa, Henry de Monfried and his wife,

Armgart.[43] Monfried has been variously described as a pirate, a smuggler, and an arms dealer. Nevertheless, the three immediately developed strong bonds that lasted for decades, and Teilhard would often visit them in their East African home during his many voyages between Asia and France. As one of Teilhard's biographer's comments: "Teilhard was so attracted to this couple that, still aboard the *Angkor*, he confessed to Armgart, 'I have full faith in Henry, in what he says about himself; but even more truly, I love you, you and him.'"[44]

On a return voyage from China, three years later, Teilhard stopped in East Africa to join Henry and Armgart for a visit, with apparently no reservations at all concerning the use of opium. According to Teilhard's biographer, Jacques Arnould, "Teilhard brought Monfried opium from China—'for his personal use.'"[45] On another occasion, Teilhard saved Monfried from arrest by local authorities in China when Monfried was trying to pick up a shipment of hashish in Chinese Turkistan.[46] It should be noted that the use of hashish and opium was widespread in China during this time, and we can assume that the European enclave of intellectuals and artists in Peking in the "roaring twenties" may likely have experimented with psychotropics such as hashish and opium, particularly as the practice was not illegal during Teilhard's years in Peking: "Ma Fuxiang [a Chinese warlord in the early twentieth century] officially prohibited opium and made it illegal in Ningxia [including Peking], but the Guominjun reversed his policy; by 1933, people from every level of society were using the drug."[47]

In such an environment, it is not beyond consideration that Teilhard may have experienced the psychotropic effects of hashish and/or opium, which would have only provided rich psychic material for self-observation and development of his ideas concerning a hyperphysics of consciousness, the noosphere and the Omega point.

The Mystical Sense

It is Teilhard's mystical sense aligned with his rigorous scientifically trained skill in observation as a geologist and paleontologist, coupled with his Jesuit training in logic, clarity, and expressive writing that gave him the ability to record his ideas so prolifically. In addition to eleven volumes of scientific publications published during his lifetime, there now exist thirteen volumes of speculative philosophy, none published prior to his death.[48]

It is clear that Teilhard as geologist/paleontologist was in an ideal position for observing, documenting, and interpreting the direct experiences of the inner life of Teilhard the contemplative priest. This integral configuration underlies the development of his "hyperphysics," his "physics of centration."[49]

After World War II, having just spent six years in relative isolation under the Japanese occupation, Teilhard gave a lecture at the French Embassy in Peking, in which he talks about the "growing importance with which leading thinkers of all denominations are beginning to attach to the phenomenon of mysticism."[50] He goes on to describe mysticism in the perception of the Omega point:

> Let us suppose that from this universal centre, this Omega point there constantly emanate radiations hitherto only perceptible to those persons whom we call "mystics." Let us further imagine that, as the sensibility or response to mysticism of the human race increases with planetisation, the awareness of Omega becomes so widespread as to warm the earth psychically.[51]

That Teilhard's understanding grew over the arc of his lifetime is evident in essays striving to express his vision, beginning in World War I and continuing until his death in 1955. In all of his essays can be detected his motivated energy to express as clearly as possible in words the framework of his understanding: "It seems to me that a whole lifetime of continual hard work would be as nothing to me, if only I could, just for one moment, give a true picture of what I see."[52] In 1922 Teilhard was awarded his doctorate, defending his thesis on mammals of the Lower Eocene (56 to 33.9 million years ago) in France.[53] According to a biographer, "The board of examiners had no hesitation in conferring on him the title of doctor, with distinction."[54] In that same year, the British psychologist Conway Lloyd Morgan (1852–1936) presented a series of radical new ideas as speaker at the Gifford Lectures, in which he extended the ideas of Henri Bergson.[55] Morgan described how an observed increase of complexity in the evolutionary process often results in discontinuous leaps with the past, rather than through a more gradual, steady process, as had been predicted by the theory of Darwinian natural selection.[56]

Lloyd Morgan's theory can be seen as a precursor to an expression of the dynamics of complexity-consciousness in Teilhard's own hyperphysics. The term "centro-complexity" had been first introduced by Teilhard to describe what he saw as a natural movement in nature during which separate elements come together, moving toward a common center and increasing complexity, moving toward higher states of consciousness. Examples might be seen in the billions of complex molecules that must come together to form a living organ such as a liver or brain, or the 10^{56} hydrogen atoms that must move together toward a common center to form what is known as a critical mass, leading to ignition and birth of a star.

The direct effect of centro-complexification, according to Teilhard, catalyzes transformation in the organization and functioning of consciousness, causing a phase shift, as when water crystallizes into ice, or transforms into steam. It is this principle of centro-complexity that drives, that initiates this catalysis.

Unfortunately, essays such as "Centrology," which develops the theory of centro-complexity in detail, were never published in Teilhard's lifetime. Conservative elements in the Catholic hierarchy made it difficult if not impossible for him to publish much of his work, in great part because the church had not yet reconciled the science of evolution with doctrinal Catholicism, and Teilhard's essays and lectures soared unchecked on a wave of evolutionary ideas. Though Teilhard was forbidden to teach, lecture, or publish outside of a narrow range of scientific material, yet his strictly scientific publications fill eleven volumes, indicating the extent of his output and providing an indication of his professional stature as a world-class paleontologist. Teilhard's books and essays on speculative philosophy and the evolution of consciousness, on the other hand, though published between 1955 and 1976, only after his death, fill another thirteen volumes.[57]

Certainly, being forbidden to publish had its effect on Teilhard. To keep him out of Paris, where the church saw his ideas as attracting too much enthusiasm among young seminarians, he was virtually banished from Paris, ordered to an assignment in China early in his career, and then banished again, to America, after the war and near the end of his life.[58] These challenges (some might say affronts) to the expression of his richest ideas, coupled perhaps with the horror and suffering he had experienced at firsthand during two world wars, all must have taken a toll on his emotional side, and must surely have contributed to his frequent bouts of despondence and depression. Pierre Leroy, his friend and colleague throughout their years of confinement in Peking, who, at twenty years Teilhard's junior, had first met Teilhard in 1928 in Paris, writes of Teilhard's bouts of depression:

> Many have rightly been struck by Père Teilhard's great optimism. He was indeed an optimist, in his attribution to the universe of a sense of direction in spite of the existence of evil and in spite of appearances [...] but how often in intimate conversation have I found him depressed and with almost no heart to carry on [...] During that period he was at times prostrated by fits of weeping, and he appeared to be on the verge of despair [...] Six years thus went by in the dispiriting atmosphere of China occupied by the Japanese and cut off from the rest of the world.[59]

Yet when Teilhard was finally able to leave China, at the war's end, he wrote, during the sea passage on his return to France, "These seven years have made me quite grey, but they have toughened me—not hardened me, I hope—interiorly."[60] He retained the passion and motivation to write extensively, particularly in his later years, and he continued the development of his observations and conclusions regarding consciousness and the dynamics of energy in an evolving universe. He himself would likely have characterized the gift of this persevering energy with the term "zest," which he defines here a 1950 essay: "By 'zest for living' or 'zest for life,' I mean here, to put it very approximately, that spiritual disposition, at once intellectual and affective, in virtue of which life, the world, and action seem to us, on the whole, luminous—interesting—appetizing."[61]

It is almost as if the restriction placed upon him by the church against publication gave him free rein to explore his ideas in essays that were freely distributed among his closest friends and many acquaintances. In spite of the censorship of the church, many unofficial copies of his writings were made, and most have been published in posthumous collections.[62] One of his most profound essays, "Centrology: An Essay in a Dialectic of Union," discussed in detail later in this book, was written in his period of isolation in Peking during Japanese wartime occupation.[63] Soon after emerging from his seclusion in China, Teilhard was deeply disappointed when the Vatican forbade him to publish what he considered to be his major work, *The Human Phenomenon*, while simultaneously refusing him permission to accept the offer of a prestigious teaching Chair at the Collège de France. Yet in spite of such opposition to his visionary understanding of the energy of consciousness, it has been noted that, "he wrote *more* religious and philosophical essays in the years 1946–1955 than during any other period of his life—his bibliography lists over ninety titles for this time."[64]

During the war, Teilhard had given the name "The Great Monad" to his conception and experience of an emerging consciousness.[65] But by 1920, during his doctoral studies, he was using the term "Anthroposphere" in referring to this thinking sphere of the planet.[66] In Paris in 1921, drawn together by similar interests, Édouard Le Roy (1870–1954) and Teilhard de Chardin met and became friends. A mathematician and philosopher by training, Le Roy immediately found in Teilhard an intellectual equal, and the two began a lifetime relationship, leading with the year to the exploration of a new concept, the noosphere.* Le Roy had studied with Henri Bergson, and had become known as his protégé; subsequently he had been appointed successor to Bergson at the Collège de France.[67] The two soon began a series of informal weekly discussions: "Punctually, at 8:30 p.m., on Wednesday evenings Teilhard would call at Le Roy's apartment in the Rue Cassette, and it was not long before the two men were thinking and speaking with a single mind."[68]

Though Le Roy was a decade older than Teilhard, their relationship appears to have been considerably more than simple mentorship; Teilhard says in a letter, "I loved him like a father, and owed him a very great debt [. . .] he gave me confidence, enlarged my mind, and served as a spokesman for my ideas, then taking shape, on 'hominization' and the 'noosphere.'"[69]

---

*The term derives from from the Greek νοῦς (*nous*: "sense," "mind," "wit") and σφαῖρα (*sphaira*: "sphere," "orb," "globe").

Over their many months of frequent discussion, the two grew so close in their philosophical thought that Le Roy would later say in one of his books: "I have so often and for so long talked over with Père Teilhard the views expressed here that neither of us can any longer pick out his own contribution."[70]

Their meetings soon included a mutual acquaintance, the brilliant writer Vladimir Ivanovich Vernadsky (1863–1945), a distinguished Russian geologist from St. Petersburg, who eventually founded the field known as biogeochemistry.[71] Vernadsky popularized his term "the biosphere" in a series of lectures at the Sorbonne during 1922–1923, frequently attended by Le Roy and Teilhard.*

Vernadsky viewed the phenomenon of life as a natural and integral part of the cosmos, and not merely some epiphenomenon. Accordingly, he professed that universal physical laws, discovered by science over a wide range of seemingly disparate fields, would eventually find continuation with fundamental principles that are the ground of life.[72]

Not widely acknowledged in the West, Vernadsky was the first to recognize the importance of life as a geological force, an idea that predates the more recent Gaia hypothesis:

> James E. Lovelock, the British inventor and the other major scientific contributor to the concept of an integrated biosphere in this century, remained unaware of Vernadsky's work until well after Lovelock framed his own Gaia hypothesis. Whereas Vernadsky's work emphasized life as a geological force, Lovelock has shown that earth has a physiology: the temperature, alkalinity, acidity, and reactive gases are modulated by life.[73]

Teilhard left Paris for China on April 6, 1923, booking inexpensive shipping routes that gave him opportunity to spend time exploring the Suez, Ceylon, Sumatra, Saigon, and Hong Kong, before arriving in Shanghai. During his time at sea, he had ample hours to think about and to observe the biosphere: "Teilhard spent his time aboard ship reading, writing, and observing nature. He liked to look at the stars at night—so

---

*The term *biosphere* had been in use since as early as 1900, popularized by the Austrian geologist Eduard Suess. Teilhard, "Centrology," 102.

clear and bright when seen from a ship far from the intruding lights of terra firma—and by day observe the state of the ocean, calm at times and stormy at others."[74]

On May 6, 1923, barely a month after departing from Marseille, Teilhard completed the essay, later called "Hominization," that sets forth his first extended exploration of the "noosphere" concept, which may be considered an outgrowth of recent discussions with Vernadsky and Le Roy in Paris.[75] In the essay, Teilhard begins by making a subtle shift from the usual Cartesian linear approach to paleontological classification toward a more spherical, three-dimensional metaphysical geometry: "We begin to understand that the most natural division of the elements of the earth would be by zones, by circles, by *spheres*."[76] In the last half of the essay, Teilhard develops his understanding of the "noosphere" concept, and in one section, "The Psychic Essence of Evolution," Teilhard says:

> It has appeared as a possible element in a sort of higher organism which might form itself [. . .] or else something (someone) exists, in which each element gradually finds, by reunion with the whole, the completion of all the savable elements that have been formed in its individuality.[77]

In this "reunion with the whole" can be seen a foreshadowing of the main theme of one of his final essays, written thirty-two years later, "The Death-Barrier and Co-Reflection," in which is described a process in which each individual human, at least the "savable elements," transcend the physical death barrier, merging with the noosphere due to "the principles of the conservation of consciousness [. . .] conceived as the luminous attainment of *a new psychological stage*."[78]

Barely a week before his own death, Teilhard concludes his "breaking the death-barrier" essay with the statement that "the interior equilibrium of what we have called the Noosphere requires the presence *perceived by individuals* of a higher pole or centre that directs, sustains and assembles the whole sheaf of our efforts."[79] The emphasis that Teilhard places on the words "perceived by individuals" can be seen here to underscore the experiential, participatory dimension of his quest to explore and to understand the dynamics of planet Earth, considering it, certainly from Vernadsky's biospheric view, as an evolving organism at every level.

But in 1924, thirty-two years before his final essay, Teilhard found himself in a state of withdrawal as he arrived in China, a somewhat banished intellectual from Paris. A friend commenting about Teilhard's state of mind at that time, wrote, "his friends noticed that he seemed to be abstracted and withdrawn."[80] Teilhard himself writes, shortly after his arrival in China, "I feel very much as though I had reached the limit of my powers: I seem somehow unable to keep things in my mind. I have the continual feeling that as far as my own life goes, the day is drawing to a close."[81]

Arriving in the Chinese city of Tientsin (today's industrial port city of Tianjin) he joined the French Paleontological Mission in China founded by his fellow priest Father Licent (whom Teilhard soon discovered to be the sole other member of the Paleontological Mission). After a two week stay in Tientsin he found himself departing on his first expedition into upper Mongolia and the mountainous Ordos Desert with his fellow priest, Licent, who himself had been exploring Mongolia for the past nine years. They were traveling to an area in which Licent had discovered fossil deposit sites from which he had previously shipped to Teilhard in Paris specimens from the Tertiary period (65 million to 2.6 million years ago).

The two priests traveled and camped for over a year in the vast silence of this area of Mongolia. Even today, in the twenty-first century, Inner Mongolia, from Gansu to Xinjiang, is considered an isolated area of China. But in an early 1924 letter Teilhard writes, "I looked over the steppes where gazelles still run about as they did in the Tertiary period or visited the yurts where the Mongols still live as they lived a thousand years ago."[82]

It was here, during this extended period of solitude in the Mongolian high desert plains, on some silent bright day or crystalline night under the canopy of stars, that Teilhard experienced a new realization, a new communion with the universe. And if we read carefully, we can even pick out expressions of these particular moments, as recorded in *The Divine Milieu* (1927), when the young priest begins to establish this connection consciously, becoming one with the energies of a divine being in the surrounding landscape:

> On some given day a man suddenly becomes conscious that he is alive to a particular perception of the divine spread everywhere about him

[. . .] It began with a particular and unique resonance which swelled each harmony, with a diffused radiance [. . .] And then, contrary to all expectation and all probability, I began to feel what was ineffably common to all things. The unity communicated itself to me by giving me the gift of grasping it. I had in fact acquired a new sense, *the sense of a new quality* or *of a new dimension*. Deeper still: a transformation had taken place for me in *the very perception of being*.[83]

The result of this transformation was a new level of written communication and communion, as his friend Pierre Leroy recounts: "It was during this expedition, in the stillness of the vast solitude of the Ordos desert, that one Easter Sunday he finished the mystical and philosophical poem, *Mass upon the Altar of the World*."

As to how Teilhard attained to this "new sense in the very perception of being," precisely what occurred to establish the connection to this "hidden power stirring in the heart of matter, glowing centre," he does not give a clue, but he personalizes it and gives it a name "the *divine milieu*," characterizing it as a sound, a single note, an "ineffably simple vibration":

Just as, at the center of the divine *milieu*, all the sounds of created being are fused, without being confused, *in a single note* which dominates and sustains them (that seraphic note, no doubt, which bewitched St. Francis), so all the powers of the soul begin to resound in response to its call; and these multiple tones, in their turn, compose themselves into a single, *ineffably simple vibration* in which all the spiritual nuances [. . .] shine forth [. . .] inexpressible and unique.[84]

And Teilhard assures us that this new sense arises from a profound interior vision: "One thing at least appears certain, that God never reveals himself to us from outside, by intrusion, but from within, by stimulation, elevation and enrichment of the human psychic current."[85] It is this *psychic current* that becomes ever more clear and omnipresent to Teilhard as if some presence was growing within him.

In 1927 under the heading "The Growth of the Divine Milieu" he writes:

> Let us therefore concentrate upon a better understanding *of the process by which the holy presence is born and grows within us*. In order to foster its progress more intelligently let us observe the birth and growth of the divine milieu, first in ourselves and then in the world that begins with us.[86]

It becomes clear to Teilhard that there is a purpose and a goal of the growth of consciousness in the divine milieu, and that is Omega, the final letter in the Greek alphabet. The letter Omega has been used for centuries in Catholic theology and liturgy to denote "the end," or "the ultimate," the goal of human cosmogenesis, the locus of transformation from human to divine. For Teilhard the scientist, the letter takes on the specific role of symbolizing the universal Center of attraction, the Omega point, symbolizing the ultimate goal of the inward, radial flow of magnetic consciousness that Teilhard is able to sense. He calls this energy process *centration*, energy that pulls inwardly toward a center, a universal Center, Omega. In his role as a Catholic priest, Teilhard saw the metaphysics of the Omega point as a harmonious solution to the problem of bringing his vision of consciousness into alignment with his employer, the Vatican, who nevertheless at that time censored *in toto* his speculative essays on the evolution of consciousness. In a moment of optimism in a 1945 essay, Teilhard writes, "Just suppose that we identify the goal of the cosmic Christ of faith with the Omega Point of science: then everything in our outlook is clarified and broadened and falls into harmony."[87]

In 1955, almost thirty years after his early experiences of the milieu in the trenches at Verdun and in the vast Mongolian desert, and just two weeks before his death on an Easter Sunday in New York, Teilhard completed "The Christic."[88] Only twenty pages long, this final essay describes his mature vision, integrating science and religion, and recapturing his earlier first experiences of the psychic movements or currents.

In the first section of the essay, "The Amorization of the Universe," we find a summary of Teilhard's many years of experiencing "the two fundamental psychic movements or currents." These two fundamental currents can be seen to reflect his distinct vision of energy as having two mutually perpendicular components (like the Cross) or movements, a tangential flow in space and time that fills the cosmos with the creative electrical radiance of resonant matter, and a second current, a magnetic radial inflow of energy resonance, of spirit returning to the source, the process of centration, of noogenesis, the evolution of consciousness, leading toward Omega.

Upon returning to Paris, after having spent eighteen months in his extended expedition to Mongolia, Teilhard resumed teaching classes at the Institut Catholique, from which he had taken a leave of absence.[89]

Teilhard gave four lectures on evolution during the winter months of 1925; and at the same time continued to develop his theory of the noosphere—a kind of cosmic envelope created by the reflection of the mind. The word was his own invention—it had come to him during the war—but the word and the idea were both adopted later by Le Roy and the Russian geologist Vernadsky, who was in Paris at the time.[90]

It had been a previous experience of the noosphere while exploring the Ordos Desert of Outer Mongolia, that had convinced him of the living existence of a noosphere.[91] Describing this peak experience, later recounted in "The Mass on the World," Teilhard wrote: "I saw the same thing as I saw long ago in the Eastern Front (which from the human point of view, was the most alive region that existed): one single operation is in process of happening in the world, and it alone can justify our action."[92]

Teilhard returned to Paris only to discover that an earlier unapproved essay on consciousness had been discovered by a Jesuit colleague in one of his desk drawers at the Institut Catholique, and subsequently forwarded to the Vatican, "where the Holy Office and the Jesuit headquarters already held a file on Teilhard."[93] Reproaching him for having dared to discuss new ways of understanding theology, the church authorities (1) insisted that he "sign a pledge to keep silent in the future," (2) permanently revoked his license to teach at the Institut Catholique, where he had been an assistant professor in geology, and (3) asked him to leave Paris.[94] Teilhard's return to China in 1926 seems to have been at a low point in his career:

> His departure had something of the aspect of a disgrace. He had been removed from Paris by the prudence of his superiors, to whom he had been denounced for propagating dubious ideas, and who feared a censure that would be equally prejudicial to the career of the young scientist and the good name of the Order. Thus he was leaving France

under a cloud for an indefinite time, and he saw the momentum of his influence broken just as it was beginning to prove fruitful.[95]

Problems with a previous essay only aggravated the situation, one of his earliest essays, "Cosmic Life," a fifty-six-page essay written at Dunkirk on the front during Easter week, April 24, 1916. When he sent it for publishing to the editors of a Jesuit periodical in Paris, *Études,* it was rejected for including such sentences as "The life of Christ mingles with the lifeblood of evolution."[96] In the rejection letter, one of the editors explained: "Your thesis is *exciting* [he used the English word, in the midst of French] and interesting to a high degree [. . .] It is a rich canvas, full of beautiful images. But it is not at all suited for our peaceful readers."[97]

The essay had then been brought to the attention of officials at the Vatican by Teilhard's Jesuit supervisor, Father Claude Chanteur, who expressed strong reservations about giving Teilhard formal admission into the Jesuit Order, but who eventually, perhaps reluctantly, allowed the thirty-seven-year-old Teilhard to take his solemn vows, on May 26, 1918.[98]

Upon his return to China in 1926, Teilhard made the best of his virtual exile, involving himself deeply in running the Jesuit museum there, and accompanying Father Licent on extensive paleontological expeditions into the vast interior of eastern Mongolia.[99] During the next twenty years, Teilhard traveled extensively between China and Paris, but during the years from 1940 to 1946 he found himself in Peking under the Japanese occupation, unable to travel. Instead, he devoted himself to perhaps his most challenging essays on the dynamics of consciousness, including "Centrology," discussed in the following section.

In May of 1946, at the end of World War II, Teilhard returned to Paris, where he tried unsuccessfully to obtain permission to publish a major work he had completed in China over a period of several years, *The Human Phenomenon.* While in Paris awaiting a reply, Teilhard began giving lectures on his latest philosophical ideas after a hiatus of twenty-two years, and soon he began to attract the interest of young Jesuit students, as well as a more secular public that had become less conservative as a result of the war. While Teilhard scrupulously avoided large venues (he had been henceforth forbidden in 1923 to give large public addresses), he soon fell into a hectic schedule alternating between intimate private conversations and semipublic discussions. During one such series of monthly discussions with several Jesuit intellectuals, Teilhard set forth two fundamental points in his philosophy, as recorded by one of the attending priests, Fr. Lejay:[100]

- God acts globally on the whole of evolution and consequently utilizes, in selective fashion, all the possibilities offered by secondary causes.
- Evil is a by-product of evolution, for there is no evolution without groping, without the intervention of chance; consequently, checks and mistakes are always possible.[101]

After the war in Europe, the topic of evil was of great interest, not only among intellectuals, but also among the general public.[102] Teilhard's characterization of evil as a part of the process of an evolutionary energy did not sit well with conservative church authorities, nor did his growing popularity among intellectual Jesuits and the public. He was offered a chair at the Collège de France but received word from the Vatican in 1950 that he was not permitted to accept the position. That same year even his close friends and sympathetic colleagues were censured: "Jesuit academics who had espoused Teilhard's ideas, among them his friend Henri de Lubac, were ordered by the Vatican to leave their positions."[103] Teilhard decided that he could no longer reside in Paris, and after securing a position in New York, Teilhard left Paris to travel, and eventually, in virtual exile, to spend the rest of his life in North America.

In 1954 Teilhard mentioned in a letter to a nephew that when he died, he wished that it might be on Easter Sunday, the quintessential day, as a Catholic, for celebrating resurrection transformation into eternal life.[104] Perhaps this can be seen as an example of real psychic precognition, for the next year, on April 10, 1955, Teilhard attended Easter services at St. Patrick's Cathedral and enjoyed the Sunday afternoon in company of his close friend Rhoda de Terra and her daughter.[105] Pierre Teilhard de Chardin died while drinking a cup of tea in the front room: "Suddenly, while standing at her window, he fell full length to the floor like a stricken tree."[106] Only a few friends attended his funeral, and only two people accompanied his body on the journey to the cemetery, sixty miles north of New York City, along the banks of the Hudson River to the Jesuit novitiate of St. Andrews-on-the-Hudson.[107] "Only Père Leroy and another priest accompanied Teilhard on his last journey. The coffin had to be laid in a temporary vault because the earth was still too frozen for a grave."[108]

Teilhard is buried near the east bank of the Hudson River, under a simple stone inscribed only with his name, dates, and "R. I. P." The small grave remains, but the seminary was sold in 1970 and is now the headquarters of the Culinary Institute of America.[109]

Part 3

•

HYPERPHYSICS OF THE NOOSPHERE

7
Hyperphysics: Speculative Physics beyond Material Science

For the past few centuries, the unquestioned materialist paradigms of science have stifled and even suppressed any scientific investigation of the spirit/soul/psyche or any other phenomena that somehow involve consciousness. Recently this is changing, but our very survival is now in question. We have reached a tipping point where materialism and materialistic science and engineering have failed us, after having plunged our world of eight billion people into the very problems our technology now cannot solve.

It is time to transform the way we view the world by working to reconcile physics with metaphysics, "natural science" with "supernatural science." What is needed is a new science that works to integrate physics with metaphysics, and new career paths for a current generation of serious explorers of the dimensions of consciousness that have formerly been off limits to the experiments of physical science. Such a new science would include efforts to integrate elements of physics and metaphysics, science and mysticism, material and psyche. Unfortunately, those who have experienced a passion and drive to explore and develop such a science have not only been ostracized and ignored, but even silenced (forbidden to publish their findings) by the authorities under which they are employed.

The work of Pierre Teilhard de Chardin (1881–1955) is a case in point. Teilhard was one of the rare few capable of bridging these two domains, physics and metaphysics, in an experiential life-long search to develop the new science that he called "hyperphysics." As a priest, mystic, and scientist, he wrote extensively and produced a model for the evolution of consciousness in the universe in scientific terms. He left behind a legacy that he hoped would forge a new mysticism, a science-based religious understanding of the dynamics of cosmos. In an optimistic note, Teilhard—then a seventy-two-year-old paleontologist, geologist, and priest—in sight of St. Helena on a passage from New York, writes: "It is with irrepressible hope that I welcome the inevitable rise of this new mysticism and anticipate its equally inevitable triumph."[1]

An Integral Approach: The New Mysticism

In spite of all the theoretical objections that would seek to discourage the belief, our minds remain invincibly persuaded that a certain very simple fundamental rule lies hidden beneath the overpowering multiplicity of events and beings: to discover and formulate this rule, we believe, would make the universe intelligible in the totality of its development.[2]

PIERRE TEILHARD DE CHARDIN

Currently it is possible to identify a scientific basis for Teilhard's energy model of consciousness. Teilhard's theories can be seen to be directly supported by the recently hypothesized electromagnetic field theory of consciousness put forward by the Cambridge genetic researcher Johnjoe McFadden,[3] the New Zealand neurobiologist Susan Pockett, and others.[4] A fact that may support Teilhard's perceived radial and tangential forces acting in the universe of consciousness, the electromagnetic field that powers all of our "devices" has two forces that are perpendicular to one another: the electric field and the magnetic field.

It is my contention that Teilhard's hyperphysics emerged as a product of multiple factors: a deep contemplative mystical sense combined with extensive scientific training, intense experience, and high intelligence, acting together in Teilhard to provide a truly integral perspective. Figure 7.1 presents a symbolic diagram of these factors in Teilhard's unique multi-perspectival consciousness.

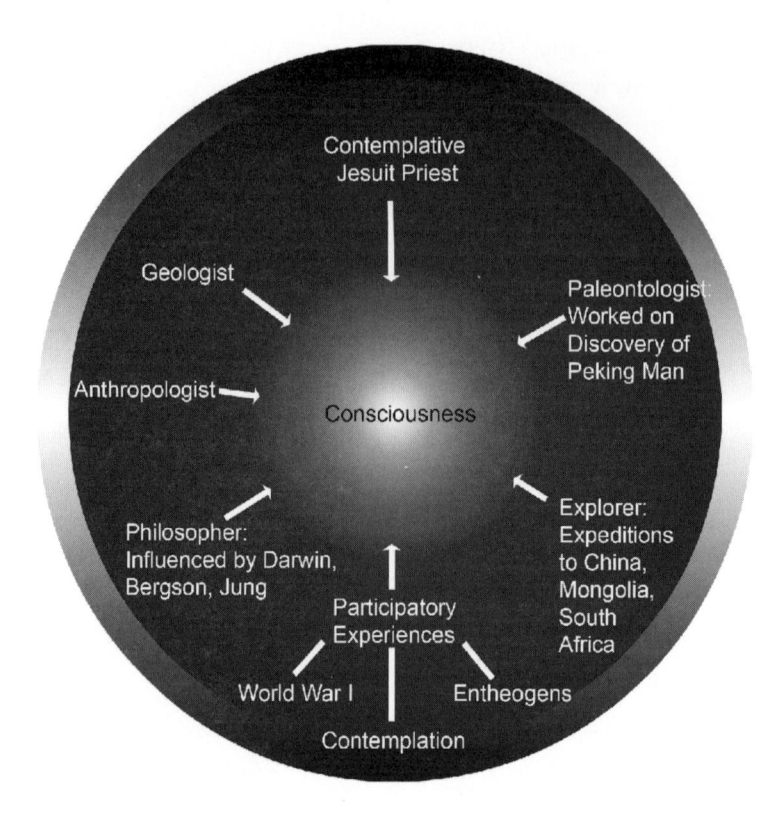

Fig. 7.1. Teilhard's multiperspectival consciousness.

It is clear in his writings that Teilhard de Chardin, the trained anthropologist, regarded his theories of hyperphysics not as philosophy or metaphysics, but as an extension of physical science. He argues for a convergence of physics and metaphysics, but not a union, and in a topological, geodesic-like metaphor, defends his hyperphysics as being distinct yet parallel to existing categories of inquiry:

> Just like meridians as they approach the pole, so science, philosophy, and religion necessarily converge in the vicinity of the whole. They converge, I repeat, but without merging, and never ceasing to attack the real from different angles and levels right to the end [. . .] It is impossible to attempt a general scientific interpretation of the universe without *seeming* to intend to explain it right to the end. But only take a closer look at it, and you will see that this hyperphysics still is not metaphysics.[5]

The History of the *Nous* and the Noosphere

One of the earliest discussions of the *nous* or *noos* may be found in the writings of the pre-Socratic Greek philosopher who introduced philosophy to Athens, Anaxagoras (500–428 BCE).[6] In developing his philosophy of the infinite interconnectedness of an infinite multitude of imperishable small parts, Anaxagoras concluded the following:

> A single overarching principle is needed to provide unity to the whole system. This principle is *nous* [. . .] "Nous" is more related to the concept of mind in the sense of the human mind or reason (though distinct from "logos," which is also sometimes translated as reason). It represents, furthermore, a kind of unity of thought a "thinking thing" in some sense.[7]

The Neoplatonists developed the concept of the *nous* even further 800 years later, and this can be found particularly in Plotinus.[8] According to Plotinus, writing in the third century CE, the original Being, the One, emanates the nous, the archetype of all manifestations in the visible world of time and space.[9] "Plotinus presents a philosophy of Unity: unity as unfathomable and transcendent, and unity as omnipresent and immanent."[10]

This Neoplatonist nous is accessible to the human mind under certain conditions, and it is what the Neoplatonists termed the *anima mundi*, or "world soul," which bridges the nous with the material world of time and space.[11] The noosphere for Teilhard, as expressed in his numerous essays, corresponds more with the Neoplatonist anima mundi.

Teilhard's concept of the noosphere is indeed part of the phenomenal world while maintaining links to the transcendent; but it is specifically associated with the planets in general, and the Earth in particular, with human consciousness evolving within a planetary sphere. Teilhard goes so far as to discuss the possibility of multiple, numerous noospheres, associated with distant planets, and speculates that there may indeed be communication between these multiple noospheres.[12]

Though there had been some controversy over the origin of the word *noosphere*, shortly before his own death Teilhard confirmed that the word was his own in a letter referring to the recent demise of his friend Édouard

Le Roy. In the letter, Teilhard writes: "I believe, so far as one can ever tell, that the word "noosphere" was my invention; but it was he [Le Roy] who launched it."[13]

In a 1951 essay, almost thirty years after first using the term, Teilhard elaborates his mature understanding of the noosphere:

> It is an amazing thing—in less than a million years the human "species" has succeeded in covering the earth: and not only spatially—on this surface that is now completely encircled mankind has completed the construction of a close network of planetary links, so successfully that a special envelope now stretches over the old biosphere. Every day this new integument grows in strength; it can be clearly recognized and distinguished in every quarter; it is provided with its own system of internal connections and communications—and for this I have for a long time proposed the name of *noosphere*.[14]

In a collection of essays, *The Biosphere and Noosphere Reader*, the editors begin their preface by characterizing the search for the noosphere: "The noosphere lies at an intersection where science and philosophy meet [. . .] an interdisciplinary domain of wide interest and high relevance that remains outside the purview of most specialists, but is of major significance for the future of humankind."[15]

Four approaches to Teilhard's concept of the noosphere are presented:

1. The noosphere is a product of the biosphere as transformed by human knowledge and action.
2. The noosphere represents an ultimate and inevitable sphere of evolution.
3. The noosphere is a manifestation of the global mind.
4. The noosphere is the mental sphere in which change and creativity are inherent although essentially unpredictable.[16]

Teilhard's Hyperphysics

Teilhard sees the current human phenomenon of consciousness to be in the early stages of what he terms a "noogenesis," a change of state in human consciousness into a more powerful union, a "joining with other centres of cosmic life to resume the work of universal synthesis on a higher scale."[17] The dynamics of this change and the architecture of consciousness itself is the subject of Teilhard's hyperphysics. Teilhard's ideas are clear, his writing style is straightforward and his logic transparent. But appreciation of his more detailed observations on the dynamics of an energy of consciousness requires careful examination of his more technical writing. Teilhard was fascinated by the phenomenon of "change of state," such as when water changes state to become ice, and in particular, of evolutionary changes of state, such as when matter changes state to become life. But his greatest interest can be seen in potential imminent changes of state in consciousness itself, both human and cosmic, as he observed here in 1937 (note in the quote that Teilhard often uses the words *spirit* and *consciousness* interchangeably): "The phenomenon of spirit [consciousness] is not therefore a sort of brief flash in the night; it reveals a gradual and systematic passage from the unconscious to the conscious, and from the conscious to the self-conscious. It is a *cosmic change of state*."[18]

Teilhard describes three critical points in the evolutionary arc of consciousness on earth, three changes of state:

- "First the appearance of life whence emerged the biosphere."
- "Then, the emergence of thought which produced the noosphere."
- "Finally, a further metamorphosis: the coming to consciousness of an Omega in the heart of the noosphere."[19]

In his numerous essays Teilhard constructs the picture of a panoramic evolutionary arc: the Earth, having experienced a change of state at the moment when life appeared, experienced another change of state as life erupted into self-reflection (thought) in the biological zenith of *Homo sapiens*. Teilhard predicts that this evolutionary arc is now moving toward yet another change of state, and is in the process of transforming human consciousness, collectively, into an even greater "complexity-consciousness,"[20] an "internal centro-complexification,"[21] both in the individual as well as the species, through the process of "centration" or "centrogenesis."[21] These terms have all been constructed and developed by Teilhard to articulate and support his theory, and can be regarded as specific to his theory of hyperphysics. We will examine and clarify these terms by focusing on three essays in which his physics of consciousness, hyperphysics, is set forth—two written in 1937, the third and most technical written in 1944—near the end of Teilhard's long seclusion in Peking under Japanese occupation:

- "The Phenomenon of Spirituality" (1937), written during an ocean voyage
- "Human Energy" (1937), written in Peking
- "Centrology" (1944), written in Peking

While major components of Teilhard's hyperphysics are presented in these three essays, additional insights into the same concepts can be found throughout his many other writings, both published and unpublished.

However, Teilhard often uses the word *spirit* when *consciousness* would appear to be more appropriate, and henceforth in quoting passages from Teilhard we will provide an alternate reading of "consciousness" via brack-

Fig. 7.2. The alchemical ouroboros. Graphic by Pelekanos (1478).

ets, where deemed appropriate, as in "the phenomenon of spirit [consciousness] [. . .] is the thing we know best in the world since we are itself."[22]

In the opening of his essay, "The Phenomenon of Spirituality," Teilhard argues that consciousness, whether a force or an energy, should be regarded as a natural, real phenomenon in the universe, worthy of study alongside other equally "real" phenomena that are taken as objects of interest in science (e.g., light, heat, electromagnetism, gravity, etc.): "Around us, bodies present various qualities: they are warm, colored, electrified, heavy. But also in certain cases they are living, conscious. Beside the phenomena of heat, light and the rest studied by physics, there is, just as real and *natural*, the *phenomenon of spirituality*."[23]

Teilhard finds it surprising that humans have never truly come to understand this spirit/consciousness in which we are all glaringly immersed:

> The phenomenon of spirit [consciousness] has rightly attracted human attention more than any other. We are coincidental with it. We feel it from within. It is the very thread of which the other phenomena are woven for us. *It is the thing we know best in the world since we are itself,* and it is for us everything. And yet we never come to an understanding concerning the nature of this fundamental element.[24]

Teilhard describes the two most conventional approaches traditionally taken in regarding the phenomenon of consciousness:

1. Religious traditions regard consciousness [spirit], in general, to be of a transcendent nature, not of this physical, spacetime world, while by contrast,
2. Modern science regards consciousness as an epiphenomenon, a unique accident in the recent evolutionary history of the planet.

Teilhard tells us that in this essay he will propose and develop an alternative to these two approaches:

> I propose in these pages to develop a third viewpoint towards which a new physical science and a new philosophy seem to be converging

in the present day: that is to say, that spirit [consciousness] is neither super-imposed nor accessory to the cosmos, but that it quite simply represents the higher state assumed in and around us by the primal and indefinable thing that we call, for want of another name, the "stuff of the universe."[25]

Teilhard tells us that the phenomenon of consciousness has been overlooked as an object of study within physical science because, at first sight, the "consciousness portion of the world presents itself in the form of discontinuous, tiny and ephemeral fragments: a bright dust of individualities," while in truth the dimensions of this consciousness ought to be taken as "the dimensions of the universe itself."[26]

But in order to see this, says Teilhard, we need to develop a new form of perception, a new sense with which we may "educate our eyes to perceiving collective realities." Teilhard predicts the development not only of a new form of "direct vision," but the emergence of a previously unsuspected psychic ability, a new sensory mode:

> Men have for long been seeking a means of immediately influencing the bodies and souls around them by their will, and of penetrating them by *a direct vision* [. . .] Nothing seems to me more vital, from the point of view of human energy, than the spontaneous appearance and, eventually, the systematic cultivation of such a "cosmic sense."[27]

Teilhard apparently had sensed evolutionary transformations both through his direct inner vision, as well as through his outer vision, he has critically observed both an internal as well as an external nature. Evolution, claims Teilhard, is often accompanied by sudden changes of state, as in water that is seen to become ice, or a solution in crystallization, change of state in "not only molecular or atomic complexity, but interiorization."[28] Teilhard perceives, both internally and externally, that in consciousness change of state follows a process of centration or compression. This can be compared with and contrasted to entropy, the movement of expansion, diffusion, and dissipation, and together they can be seen as "two fundamental cosmic movements [. . .] which we can grasp experientially."[29] He describes these two contrary movements as the concurrent movement of energy in

Entropy (Diffusion)   Centration (Vitalization)

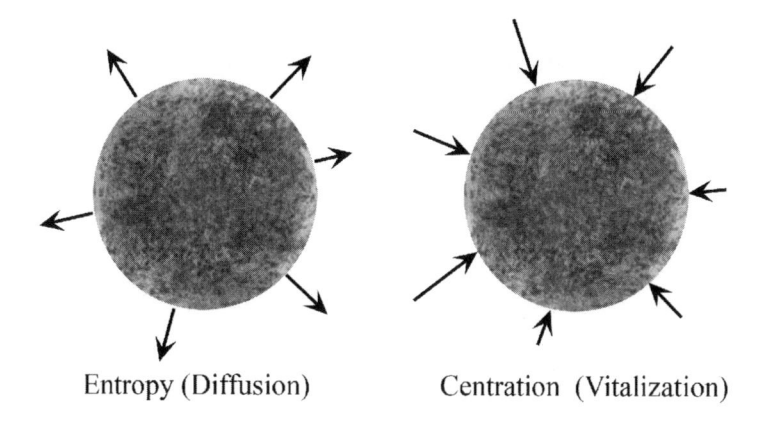

Fig. 7.3. Entropy and centration.

two directions, the "vitalization" and the "dissipation" of energy and says that they "are merely the opposite poles of a single cosmic event."[30] Figure 7.3 contrasts the two movements of entropy and centration.

In words that parallel David Bohm's description of an ongoing process of enfolding and unfolding between an implicate and an explicate order, Teilhard describes "the inward furling from which consciousness is born [. . .] around a centre [. . .] the All becoming self-reflective upon a single consciousness."[31]

At this point he brings up a theme that will arise repeatedly in his later essays, the transcendence of death by the individual personality.* Teilhard says, "to become super-conscious the fragmented building blocks of man must unite itself with others,"[32] but without losing personalities previously acquired, without losing information. Recall that Teilhard had sensed this phenomenon as a totalization of multiple centres† of consciousness, in 1916 at the Front.

---

*Shortly before his death in 1955, Teilhard wrote his final essay, "The Death-Barrier and Co-Reflection, or the Imminent Awakening of Human Consciousness to the Sense of Its Irreversibility."

†Note that the British spelling "centre" is used here and throughout textual discussion in this chapter, not only because it is in accordance with the spelling found in all published translations of Teilhard's work into English, but more specifically because in the context of Teilhard's metaphysics, the word *centre* is used to designate a "center of consciousness," rather than used simply as an adjective, or a location designator.

He goes on to state that humans in general seem to have lost the "faculty of totalization," with the exception of a few mystics, who have been able to experience union by dissolution, much as salt in the ocean. But it is union by differentiation that interests Teilhard, not union by dissolution.

> We can see a justification ahead for our hope of a personal immortality [. . .] without becoming confused with one another [. . .] to complete ourselves we must pass into a greater than ourselves [. . .] In this convergent universe, all the lower centres unite, but by inclusion in a more powerful centre. Therefore. they are all preserved and completed by joining together.[33]

Teilhard now turns to the implications of such a theory for morality, in a section he calls "Moral Applications." He describes two categories of morality, the "Morality of Balance" and the "Morality of Movement."[34]

Morality of Balance vs. Morality of Movement

Teilhard describes two types of morality, the morality of balance and the morality of movement. The old morality is the morality of balance, an attempt at homeostasis, a "morality that arose largely as an empirical defense of the individual and society."[35]

Teilhard goes on to say that, "Morality has till now been principally understood as a fixed system of rights and duties intended to establish a static equilibrium."[36] However in light of the modern discovery of the evolutionary nature of everything in the universe, the human being must be seen as "an element destined to complete himself cosmically in a higher consciousness in process of formation," and thus the need for a new morality, a morality of growth (movement), one that will foster and catalyze evolutionary change, a growth into a new formation of being. Teilhard says that new times and a new understanding of the trajectory of life and consciousness implies that while the "moralist was up to now a jurist, or a tight-rope walker," the moralist of the future must "become the technician and engineer of the spiritual [consciousness] energies of the world."[37] He says that for those who see the development of consciousness "as *the* essential phenomenon of nature [. . .] morality is consequently nothing less than the higher development of mechanics and biology. The world is ultimately constructed by moral forces."[38] Having argued the urgent requirement for a new morality, a morality of growth, Teilhard sets forth "three rules that clearly modify or complete the idea we have of goodness and perfection":[39]

- Good "is what makes for the growth of spirit [consciousness]."
- Good is everything that brings "growth of consciousness to the world."
- "Finally, best is what assures their highest development to the spiritual powers [consciousness] of the earth."[40]

Teilhard summarizes, "many things seemed to be forbidden by the morality of balance which become virtually permitted or even obligatory by the morality of movement." For example, in following a morality of balance, as long as we follow society's rules, we are permitted to waste our lives in any frivolous pursuit (i.e., in sheer entertainment); whereas under a morality of movement such things as research through experimentation with psychotropic drugs and participatory exploration of multiple religions would likely be permissible. With disregard of likely disapproval by Vatican censors, Teilhard urges a new human morality of growth (see table below) that "will forbid a neutral and 'inoffensive' existence, and compel him strenuously to free his *autonomy* and *personality* to the utmost," and he urges us "to try everything and to force everything in the direction of the greatest consciousness."[41]

Teilhard's Two Moralities
(Static Balance vs Evolutionary Growth)*

Morality of Balance	Morality of Growth
Homeostasis; closed.	Evolutionary movement; open.
Fixed rules, rights, and duties to sustain the present.	Whatever fosters growth of consciousness for the future.
Love is subordinate to procreation.	Love gives incalculable spiritual power.
The old moralities of balance are static, powerless to govern the earth.	What is needed is a new morality of movement, of growth.

*Adapted from Teilhard, "The Phenomenon of Spirituality," 105–10.

One can see from the table above why the more conservative church authorities might have had problem with these ideas, but Teilhard is an unapologetic explorer (and a mystic), who states unequivocally:

> The boldest mariners of tomorrow will set out to explore and humanize the mysterious ocean of moral energies [. . .] our goal is to try everything and to force everything in the direction of the greatest consciousness [. . .] ever since its beginnings life has been a groping, an adventure, a risk, and in general and highest law of morality: to limit force is sin.[42]

At the conclusion of his essay, Teilhard urges us to "situate the stuff of the universe in consciousness [. . .] and to see nature as the development of this same consciousness." He regards the idea of a cosmos as "a moving towards personality," and he concludes with a statement that once more would likely win him no affection among more conservative elements in the Vatican: "This is the origin of the present crisis in morality [. . .] a powerlessness of (old) moralities of balance to govern the earth. It is necessary for the religions to change themselves [. . .] what we are all more or less lacking at this moment is a new definition of holiness."[43]

Holoflux Theory and Teilhard's "Spirit"

At the heart of Teilhard's essay on morality, Teilhard introduces a major hypothesis in a key paragraph that is essential to understanding his hyperphysics, and that accords well with the quantum physics of David Bohm, in which the cosmos is seen both as simultaneously unfolding and enfolding:

> Everything that happens in the world, we would say, suggests that the unique centre of consciousness around which the universe is furling could only be formed gradually, through a series of diminishing concentric spheres, each of which engenders the next; each sphere being moreover formed of elementary centres charged with a consciousness that increases as their radius diminishes. By means of

this mechanism each newly appearing sphere is charged in its turn with the consciousness developed in the preceding spheres, carries it to a degree higher in each of the elementary centres that compose it, and transmits it a little further on toward the centre of total convergence.[44]

This description can be seen as congruent with the theory of holospheres, in which smaller dimensions converge to the limit found at the Planck holosphere, at which point begins the implicate order.

Teilhard concludes this section by telling us that "the final centre of the whole system appears at the end both as the final sphere and as the centre of all the centres spread over this final sphere."[45] This "centre of all centres" can be understood as Bohm's implicate order, transcending spacetime. Teilhard also refers to this center as "a quantum of consciousness," and tells us that each degree of consciousness at a given moment only exists as "an introduction to a higher consciousness," and that this general process is irresistible and irreversible.[46]

If we accept this hypothetical model, says Teilhard, then we are led to two conclusions for the present and for the future:

1. The source of all our difficulties in understanding matter is that it is habitually regarded as inanimate.[47]
2. We are moving toward a higher state of general consciousness . . . other spheres must exist in the future and, inevitably, there exists a supreme center in which all the personal energy represented by human consciousness must be gathered and "super-personalized."[48]

How might we understand Teilhard's use of the term *spirit* in terms of modern consciousness theory? It is evident that Teilhard's "spirit" appears to be equivalent to holoflux energy as it manifests within Bohm's implicate order. In the holoflux theory,[49] spirit has a *nonlocal* locus embedded within a plenum of Planck-length spherical centers at the granular bottom of spacetime, everywhere. Conversely, what is termed consciousness is localized in spacetime, manifesting as expanding flux in detectable fields of electromagnetic waves, as illustrated in figure 7.4.

SPIRIT   **CONSCIOUSNESS**
NON-LOCAL   LOCAL

Fig. 7.4. Nonlocal spirit versus local consciousness.

Yet they are in relationship, they both exist as part of Bohm's "Wholeness," and there exists a direct connection between spirit and consciousness through the phenomenon of frequency resonance (the connection can be seen in the mathematics of Fourier transforms, discussed in chapter 7).

It is useful to go topologically further into the holoflux analogy. Imagine the communal locus of Teilhard's "spirit" as it is found within, at the very center of every "point" within spacetime. The holoflux process *is* the implicate order, is *one with* the implicate order; "spirit," as implicate order holoflux, has the advantage of being self-superpositioned, fully interconnected, transcending the limits of time and space and can be identified in electrical engineering terms as the frequency domain.

Let us now move outward in scale, bridging the transition zone, the isospheric shell that divides the implicate order from the explicate order. Here we see holoflux emerging from the implicate order as it transforms into spacetime energy, flaring forth as waves of spherically vibrating electromagnetic energy. These waves of energy emerge everywhere into spacetime from a holoplenum of Planck diameter isospheres. Each isosphere can be seen to encapsulate the entire implicate order, within which the infinity of frequencies from all time and all space are eternally enfolding into a hyper-harmonic flux.

In terms of Bohmian holoflux theory, the first approach to consciousness, the religious, is focused almost exclusively upon an implicate domain, energy as a transcendent flux, and a focus that generally ignores or rejects as unreal the spacetime explicate domain; conversely, in the second approach to consciousness, the modern scientific, the focus is upon spacetime explicate mode phenomena, completely ignoring the possible reality of a non-spacetime domain. Teilhard proposes an alternative to these two, seemingly mutually exclusive, approaches:

I propose in these pages to develop a third viewpoint towards which a new physical science and a new philosophy seem to be converging at the present day: that is to say that spirit is neither super-imposed nor accessory to the cosmos, but that it quite simply represents the higher state assumed in and around us by the primal and indefinable thing that we call, for want of a better name, the "stuff of the universe." Nothing more; and also nothing less. Spirit is neither a meta-phenomenon nor an epi-phenomenon; it is *the* phenomenon.[50]

Evidence of Teilhard's "spirit," or "*the phenomenon*," can be seen in the "Holy Spirit," from Teilhard's Catholic teaching of the Holy Trinity. It is in accord with the sub-quantum holoflux model that views the Whole as one single energy, processing within and between two primary domains, a spacetime domain, and a spectral domain.

Figure 7.5 shows a complete Planck diagram of the Holy Trinity highlighting Teilhard's distinction between *tangential consciousness* and *axial consciousness* mapped by Bohm's distinction between a nonlocal (implicate) order, and a local (explicate) order in space and time.

The Father → **Axial Consciousness** NON-LOCAL IMPLICATE *Spectral* ORDER **Holoflux Energy** (frequency-phase)

The Holy Spirit → *Fourier Transform*

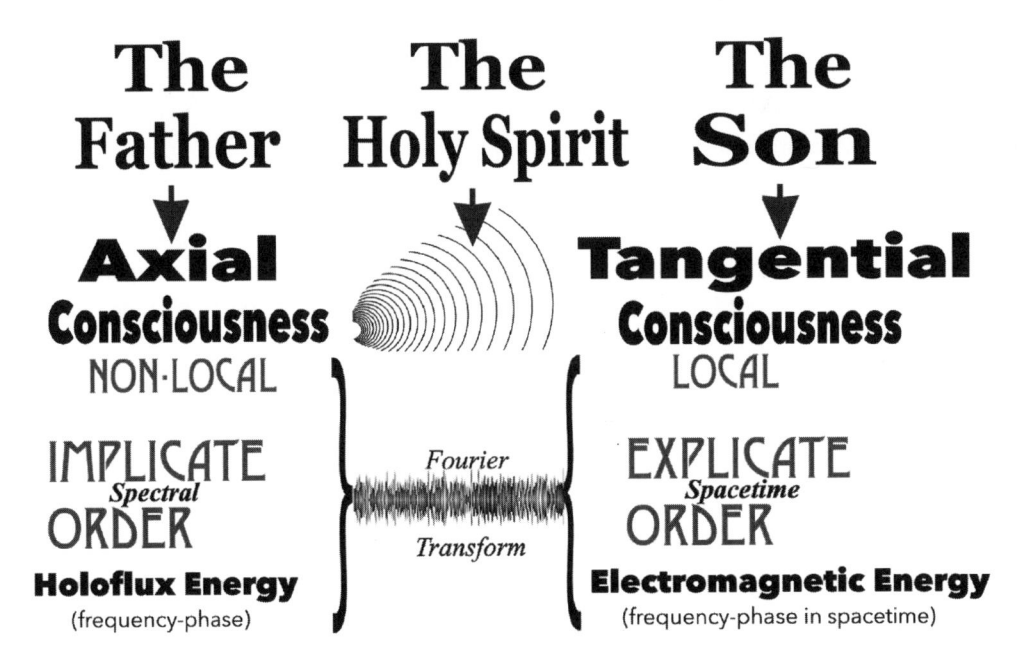

The Son → **Tangential Consciousness** LOCAL EXPLICATE *Spacetime* ORDER **Electromagnetic Energy** (frequency-phase in spacetime)

Fig. 7.5. A Planck diagram of Teilhard's Trinity.

The figure identifies "spirit" in the center, bridging the nonlocal implicate order and the local explicate order. Between the two orders that make up the Whole we see the energy processes of transformation named by mystics as a holy spirit, a mysterious "dark energy" that physicists are currently trying to understand.[51] Conversely, "consciousness" can be seen on the right, in the spacetime region, manifesting as the energy of consciousness in time and in space and also identified here as electromagnetic energy.

A Christian approach to the diagram would consider God the Father as hypostasis of the nonlocal implicate order, the Holy Spirit as hypostasis of continuous two-way Fourier transforms, and the Son as the manifestation of consciousness within space and time.

In the first section of Teilhard's essay, "The Amorization of the Universe," we find a summary of his many years of experiencing "the two fundamental psychic movements or currents." These two fundamental currents can be seen to reflect his distinct vision of energy as having two mutually perpendicular components (like the Cross) or movements, a tangential flow in space and time that fills the cosmos with the creative electrical radiance of resonant matter, and a second current, a magnetic radial inflow of energy resonance, of spirit returning to the source, the process of centration, of noogenesis, of love that pulls toward Omega.

This is the tangential seen here as the very flux of collective thought on planet Earth, an experience of the matrix-like web of human consciousness active in the biosphere.

> On the one side, [says Teilhard] there has been the irresistible convergence of my individual thought with every other thinking being on earth [. . .] [and] a flux, at once physical and psychic. And on the other side, [. . .] a centration of my own small ego [. . .] a sort of Other who could be even more I than I am myself [. . .] a Presence so intimate that it could not satisfy itself or satisfy me, without being by nature universal.[52]

Here is expressed the magnetically radial component of consciousness, the energy of centration, pulling inwardly toward a center that, perhaps beyond the event horizon of spacetime, is also the universal Center, the Omega, the ultimate personalization of human cosmogenesis, fully personalized.

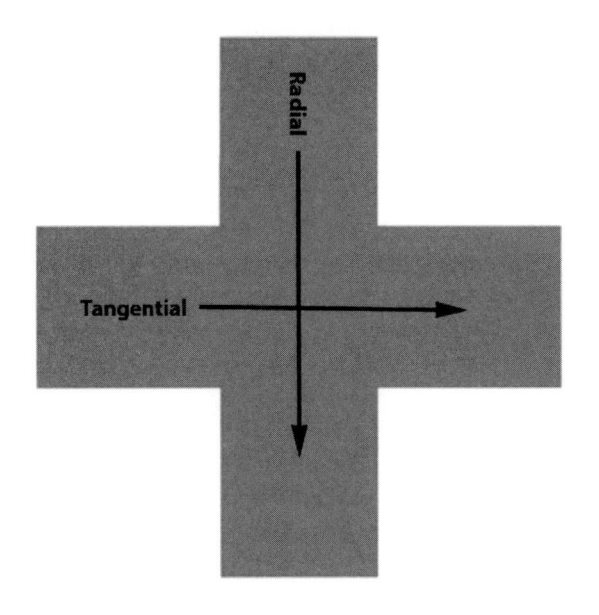

Fig. 7.6. Teilhard's tangential and radial tendencies of consciousness.

The Cross, the Trinity, and the Centric

In 1950 Teilhard describes his vision of the tangential and radial tendencies of consciousness (fig. 7.6). He begins by reiterating a recurrent theme in his writing, talking about the "two tendencies" between which there is a "heightening of the antagonism between the 'tangential' forces that make us dependent upon one another," and "the 'radial' aspirations that urge us towards attaining the incommunicable core of our own person."[53] These two tendencies, seen as vectors of the energies of consciousness standing at right angles to one another, form a cross, the intersection of which is our own center of humanized consciousness.

This "heightening of antagonism" Teilhard describes can be seen so clearly today in the seemingly abject polarity between the materialist geopolitical, corporate forces of the *tangential* as contrasted with the increasingly spiritualized *radial* forces of humanistic liberal thinkers, altruists, mystics, and the religious. It can be seen in the political polarization within our own government and clearly within the polarization within countries across the world in recent years.

Teilhard would understand this polarization in the twenty-first century and say that under irresistible pressure, our planet "is contracting upon itself."

The Centric

Ω

The Christic

The Human

The Cosmic

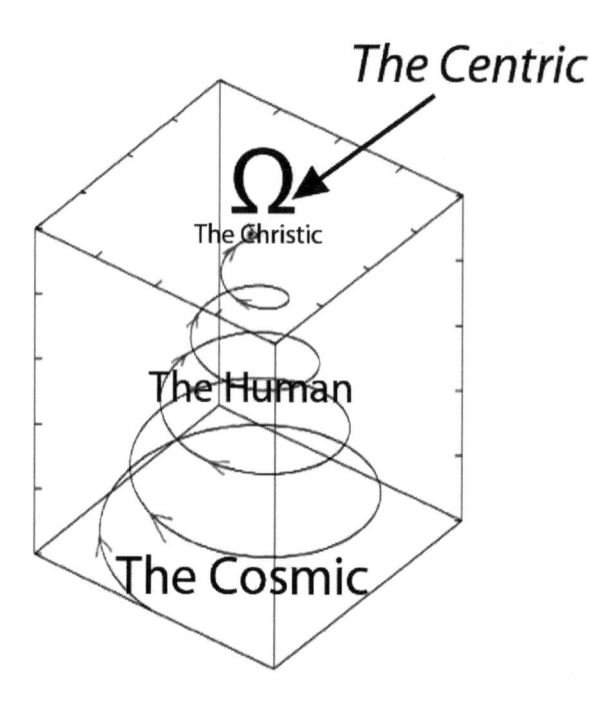

Fig. 7.7. Teilhard's Spiral: Cosmic, Human, and Omega.

But in spite of these deeply felt antagonisms, Teilhard maintains a great reservoir of hope, based upon what he foresees as resolution in a spiral of complex centrification (fig. 7.7) toward Omega, the Centric: "Here we move into what is indeed a remarkable, an astonishing region where the Cosmic, the Human and the Christic meet and so open up a new domain, *the Centric*; and there the manifold oppositions which constitute the unhappiness and anxieties of our life begin to disappear."[55]

The Noosphere and Omega

Thus in the broad arc of the evolution (and humanization) of consciousness in the universe, Teilhard's final vision is a model that gives us a way of understanding how the forces of centro-complexification have led the energies of *the Cosmic*, after 14.6 billion years, into and through convergence into *the Human*, which is even now further centro-complexifying through the humanization of an isosphere, the noosphere.

And at the center of this noosphere, powered by the Centric (fig. 7.8) is *the Omega,* into which inexorably are being drawn the cosmic and the human at the threshold.

Hyperphysics: Speculative Physics beyond Material Science • 169

The Cosmic *The Human* Ω *The Christic*

The Centric

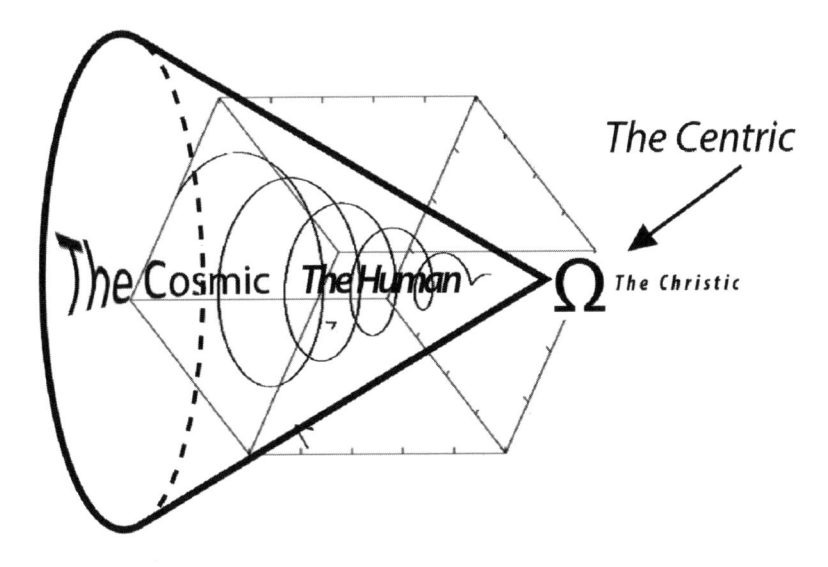

Fig. 7.8. Noogenesis and the noosphere.

At the age of sixty-five, while still in Peking, Teilhard writes that religious thinking "cannot develop except traditionally, collectively, and 'phyletically.'"[56] Such statements in his writings were widely condemned by his Catholic superiors, though later in the same essay, perhaps to mollify the Vatican censors, he suggests that if "we identify the cosmic Christ of faith with the Omega Point of science, then everything in our outlook can be clarified."

In the same essay, Teilhard has married his educationally derived scientific thinking with his religious thinking. Having established a point of commonality, he ends his essay with the section, "A New Mystical Orientation: The Love of Evolution," in which he discusses "the heart, with all that the word implies of vital and dynamic fullness."

Omega Point of Science = **Cosmic Christ of Faith**

Fig. 7.9. Teilhard's identity of Omega point with Cosmic Christ.

Cosmogenesis = Christogenesis

Fig. 7.10. From Teilhard's Catholic view, cosmogenesis equals Christogenesis.

Now, having established the common mathematical point of convergence, Teilhard reenergizes the universe of cosmogenesis with the amorization of Christogenesis.

He says "we now find that it is becoming not only possible but *imperative* literally to *love* evolution."[57] It is through this amorization of a universe and of a process that had previously grown cold through a sterilizing vision of science after Newton, that we will be able finally "to communicate, to 'super-communicate,' with him through all the height, the breadth, the depths and the multiplicity of the organic powers of space and time,"[58] and finally on the last page of his essay, summing up the "The Love of Evolution," Teilhard writes: "Love, in consequence, is undoubtedly the single higher form towards which, as they are transformed, all the other sorts of spiritual energy converge—as one might expect in a universe built on the plane of union and by the forces of union."[59]

8

The Energetic Noosphere

Energy: Axial and Tangential

Energy is the central element in Teilhard's technical modeling of the cosmos. He says that while "in metaphysics the notion of being can be defined with a precision that is geometric," things are not so clear in physics, where the notion of energy is "still open to all sorts of possible corrections or improvements."[1] Teilhard's essays on the energy of consciousness, spanning four decades, systematically introduce a coherent range of such corrections and improvements. In the last page of his essay, "Activation of Energy," Teilhard states, "there are two different energies one axial, increasing, and irreversible, and the other peripheral or tangential, constant, and reversible: and these two energies are linked together in 'arrangement.'"[2] Thus Teilhard's hyperphysics posits two modes, domains, or dimensions of energy, not only of a *tangential component* of energy that operates within spacetime dimensions, and which is measured and explored by modern physics, but also a *radial* or *axial component* of energy. It is this axial energy that provides the direct link with the center termed Omega by Teilhard, which guides, informs, and maintains the evolutionary process throughout the spacetime cosmos.[3] He describes this radial component of energy as "a new dimensional zone" that brings with it "new properties," and he describes how increasing centration along the radial component leads to increasing states of "complexity-consciousness."[4] As Teilhard says here: "Science in its present reconstructions of the world fails to grasp an essential factor, or, to be more exact, an entire dimension of the universe

[. . .] all we need to do is to take the inside of things into account at the same time as the outside."[5]

Energy, for Teilhard, is not simply regarded as a mathematical abstraction. He views energy as the matrix of consciousness, the driver of evolution, and as a living, communicating radiation or flux. For Teilhard energy is "a true 'transcosmic' radiation for which the organisms [. . .] would seem to be precisely the naturally provided receivers."[6]

Teilhard is critical of the one-dimensional approach to energy taken by contemporary research. He asks, "What is the relationship between this interior energy [. . .] and the goddess of energy worshipped by physicists?"[7] His answer is that there are two fundamental categories or modes of energy and implies that physicists deal with but one mode. In his own words, "We still persist in regarding the physical as constituting the 'true' phenomenon in the universe, and the psychic as a sort of epiphenomenon."[8]

He also describes these two components of energy in physical and psychic terms: "*physical energy* being no more than *materialized psychic energy*,"[9] but he is not able to posit a mathematical or physical relationship between these two dimensions, other than to express the hope that "there must surely be some hidden relationship which links them together in their development."[10]

A Thinking Earth: The Noosphere

Despite clerical resistance to his ideas, Teilhard continued to be fascinated by what he saw as the emerging evolution of a collective human consciousness upon planet Earth, the emergence of a "thinking Earth," a phenomenon that he had directly intuited in his intense experiences at the front in 1917. He continued his dual work in the fields of paleontology and speculative philosophy; for example, in January 1923 he finished an essay, "Pantheism and Christianity," only to publish two months later "Paleontology and the Appearance of Man."[11]

His intense life experiences led Teilhard to the perception of an emerging planetary consciousness, what he termed the noosphere,* which after

---

*"Over and above the biosphere there is a *noosphere*." Teilhard, *The Human Phenomenon*, 124. *Noosphere* from the Greek νοῦς (*nous*: "sense, mind, wit") and σφαῖρα (*sphaira*: "sphere, orb, globe").

his death has been conceptualized as *"an ultimate and inevitable sphere of evolution [. . .] a scientific approach with a bridge to religion."*[12]

Locating the Noosphere

At this point we speculate as to where in the physical spacetime universe the noosphere might be found. To find the noosphere, let us try a thought experiment and build a likely image of it. Picture in your mind the geometry of planet Earth. Imagine the heat, approximately 13,000°C in the central core.[13] Place your consciousness at the absolute geometric-gravitational central point of the planetary core. Now, begin to slowly move (or rise) outward along a radial line toward the cold of space, noting the temperature drop as you move away from the center of the planet—and stop at the moment you arrive at the temperature 98.6°F, the average human core temperature.

By repeating the above procedure multiple times, with many different radii moving at various angular separations away from the core, a three-dimensional surface mapping, like a mathematical brane or Teilhard's isosphere, will begin to emerge—an infrared energy isosphere to which each human being is linked through an identical resonance frequency.*

The shape of this isosphere will likely be highly organic and fractal in appearance, sometimes hovering above the ground on thermoclimes where the "ambient temperature" reaches 98.6°F, while in arctic regions and the oceans it will be located hundreds of feet below the surface of ice or water.

But the noosphere is more than simply a dynamic location on the surface of an isosphere at (or above or below) the rocky surface of the earth. It is energy at the same frequency band as the human body, which has been said to generate (and broadcasts) approximately 1.3 watts of radiant power with each heartbeat.[14] While we normally think of each heartbeat as simply a pushing of blood through the arteries, it is also radiantly generating infrared electromagnetic energy (the infrared being a range of the spectrum that we often hear dismissively described as "heat").

How might this information be used to substantiate Teilhard's vision of the reality of the noosphere that would manifest in some planetary

---

*A brane is a geometrical boundary of higher dimensional spaces. This concept is used in contemporary superstring theory and M-theory; see Susskind, *The Black Hole War*.

Human population since 1800 in billions

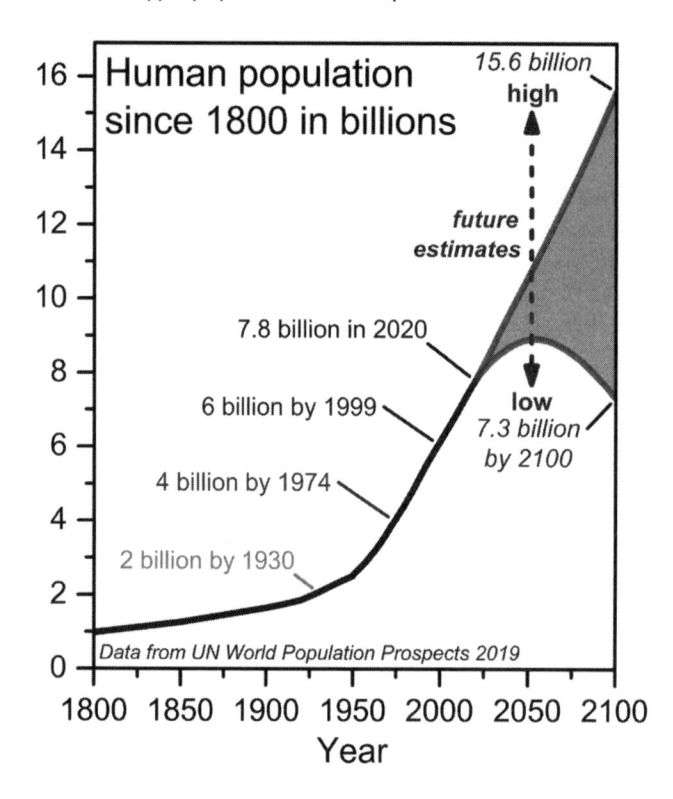

- 15.6 billion high
- future estimates
- 7.8 billion in 2020
- 6 billion by 1999
- low 7.3 billion by 2100
- 4 billion by 1974
- 2 billion by 1930
- Data from UN World Population Prospects 2019

Fig. 8.1. Human population of the Earth since 1800.

energy of consciousness? A chart of global population growth (fig. 8.1.) indicates that there are currently approximately eight billion human beings living on the planet.

Accordingly, we can multiply eight billion humans by the average of 1.3 watts of radiation per human to find the current amount of energy being broadcast by all human hearts: this calculation gives us more than nine gigawatts (10,400,000,000 watts). As can be seen in the table on the next page, this amount well surpasses the output of the most powerful radio transmitter in the world, at one million watts, even greater than the energy output of the Three Mile Island nuclear power plant's occasional maximum output.

It is possible that the nine gigawatts of electromagnetic energy being continuously broadcast by our collective heartbeats may be taking part in a vast energetic interactive resonance with Gaia. Our own collective energy, which transmits in the far infrared in the 10-micron wavelength range (predicted by Wien's law for our body temperature range) is the part of the geomagnetosphere that is us, the noosphere (the "us" sphere).

Radiant Energy Power Outputs Compared

Source of Energy	Power
Radiant electromagnetic power output of one human heart	1.33 watts
Power output of most powerful radio transmitter on planet[15]	2,500,000 watts
Maximum power output of Three Mile Island nuclear reactor[16]	873,000,000 watts
Combined power output of 8 billion human heartbeats	10,773,000,000 watts

Evidence of direct interaction of the electromagnetic energy of the geomagnetosphere with human consciousness can be viewed in figure 8.2, taken from a paper analyzing the events of September 11, 2001.

Figure 8.2 shows a chart recording daily data from Geostationary Operational Environmental Satellites and weather satellites in geosynchronous orbit over the United States in the days before, during, and after the September 11, 2001, terrorist attacks. Continuous readings show a marked peak on September 11, 2001, followed by several days of marked disruption in the observed diurnal rhythm of the geomagnetosphere.[17]

In the conclusion of the paper, the authors state, "The study [. . .] supports the hypothesis that humanity is connected via a global field."[18]

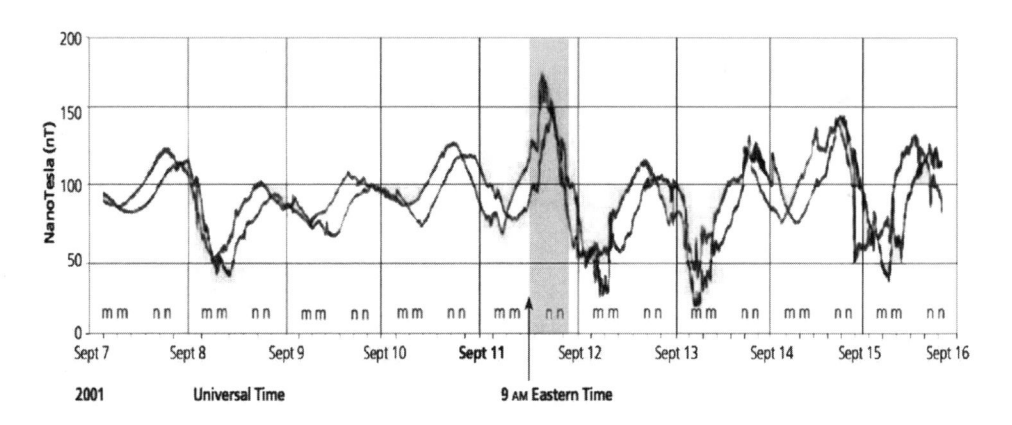

Fig. 8.2. Evidence of a coherent planetary standing wave.
Reprinted with permission from HeartMath Institute. Image from McCraty, Deyhle, and Children, "The Global Coherence Initiative," 75, fig. 8.2.

Perhaps the same hypothetical "global field" of radiation can be seen in the one Teilhard describes in a 1953 essay, "A Sequel to the Problem of Human Origins":

> Our minds cannot resist the inevitable conclusion that were we, by chance, to possess plates that were sensitive to the specific radiation of the "noospheres" scattered throughout space, it would be *practically certain* that what we saw registered on them would be a cloud of thinking stars.[19]

Human Energy

In an essay written in Peking in 1937, "Human Energy," Teilhard describes three forms of energy, and implies that modern science only considers the first two, ignoring the third; these three he identifies as:

- incorporated energy
- controlled energy
- spiritualized [conscious] energy.

Incorporated energy manifests in rocks, crystals, neurons, and so on. Controlled energy is that generated by humans and used to power human devices thermodynamically and electrically. Energy of the third kind, Teilhard's "spiritualized energy," or we would say, the "energy of consciousness," is the primary subject of his essay, "Human Energy."[20]

In this essay Teilhard proposes that each human "represents a cosmic nucleus [. . .] radiating around it waves of organization and excitation within matter."[21] Teilhard immediately proposes, based it seems upon his own experience, that these radiations can be perceived by human beings, and he makes reference to the need for development of a special psychic mode of perception: "This perception of a natural psychic unity higher than our 'souls' requires, *as I know from experience*, a special quality and training in the observer. Like all broad scientific perspectives it is the product of a prolonged reflexion, leading to the discovery of *a deep cosmic sense*."[22]

Teilhard warns that discovery of this deep cosmic sense is a matter of perception, of tuning, of intent. He says that we are like a cell that can see nothing but other cells, but that there are more complex configurations of being if we only can learn how to join with them. He says that "the thoughts of individuals [. . .] form from the linked multiplicity, a single spirit of the earth," and that he sees humanity continuing to evolve "in the direction of a decisive expansion of our ancient powers reinforced by the acquisition of certain new faculties of consciousness."[23] Teilhard emphasizes that this growth is not a work of random chance, but that it unfolds within a universe that is alive with an energy that is also synonymous with the mystery we call love or allurement: "Love, by the boundless possibilities of intuition and communication it contains, penetrates the unknown; it will take its place in the mysterious future, with the group of new faculties and consciousnesses that is awaiting us."[24]

Here he expresses a consideration missing in most physical descriptions of energy, the category of "love," something that Teilhard includes as perhaps the most real, fundamental manifestation of energy in his hyperphysical theories. As early as 1931, in "The Spirit of the Earth," Teilhard had referred to an energy of consciousness, of sensation, of love, as manifesting in a spectrum (much as Jung, in 1946, used the imagery of the spectrum to characterize the energy of the psyche).[25] Teilhard says that "Love is a sacred reserve of energy; it is the blood of spiritual evolution."[26]

> Hominized love is distinct from all other love, because the "spectrum" of its warm and penetrating light is marvelously enriched. No longer only a unique and periodic attraction for purposes of material fertility; but an unbounded and continuous possibility of contact between minds rather than bodies; the play of countless subtle antennae seeking one another.[27]

Teilhard sees this organic love-consciousness energy growing more complex and changing through some natural evolutionary process, currently unknown, but he is confident that there will be an eventual mastery and understanding of this same phenomenon (conscious love) in terms of physics.

Accordingly, he stresses the importance of those engaged in scientific research to turn their focus upon the human phenomenon of consciousness.

He is hopeful, telling us with conviction that "physics will surely isolate and master the secret that lies at the heart of metaphysics," and will accelerate this evolution toward the emergence of a new cosmic sense: "Nothing seems to me more vital, from the point of view of human energy, than the spontaneous appearance and, eventually, the systematic cultivation of such a 'cosmic sense.'"[28]

One in Many: The Noosphere

In his essay, entitled "Organization of Total Human Energy: The Common Human Soul," Teilhard discusses his concept of the noosphere, and sees in the process of "raising men to the explicit perception of their 'molecular' nature" that the possibility opens for them to "cease to be closed individuals, to become parts [. . .] to be integrated in the total energy of the noosphere."[29] But Teilhard is at pains to reassure the reader that this does not imply the loss of individuality, and he points out that individual human souls (the quanta of consciousness) are not like gas molecules, "anonymous and interchangeable corpuscles," but that the formation of the noosphere requires, on the contrary, a "maximum of personality" to be manifest through each human individual sub-contribution: "The utility of each nucleus of human energy in relation to the whole depends on what is unique in the achievement of each."[30]

Assuming that consciousness is evolving, and the material universe is evolving, Teilhard wonders where might the energy come from that guides the coalescing centro-complexity into such exquisite configurations, and he wonders what might be the nature of energy within this evolution (i.e., what is "informing" and "powering" this evolutionary process)? His answer can be found in a section he calls, "The Maintenance of Human Energy and 'The Cosmic Point Omega.'" He explains that such energy is axial, that it "is found to be fed by a particular current" flowing from the center, the Omega point, which, he calls a "tension of consciousness."[31]

Here Teilhard makes a diversion into a subject he repeatedly brings up, the continuation of the centers of personality after physical death. He says that: "Reflective action and the expectation of total disappearance are *cosmically incompatible* [. . .] death leaves some part of ourselves in some way, to which we can turn with devotion and interest, as to a portion of the absolute [. . .] as *imperishable*."[32]

He points out that cosmic evolution is a work of "*personal* nature," and that we are each a unique "centre of personal stuff totalizing, in itself, the essence of our personalities . . . the universal centre of attraction"; at this point he begins to discuss the "centre of psychic convergence," the noosphere and brings up the image of the sphere.*[33]

What are the implications in Teilhard's statement, "the totality of a sphere is just as present in its centre?" We could say that it implies a sharing of information storage between the isosphere in spacetime and the implicate order at the center (Wheeler's qubits of conscious information spread over the surface of a sphere, would be seen here to be in resonance with the implicate center). Next Teilhard poses a question whose solution appears to corroborate the concept of an implicate order within the quantum holosphere, and he concludes with his famous dictum, "union differentiates":

> Now, why should it be strange for the universe to have a centre, that is to say to collect itself to the same degree in a single consciousness, if its totality is already partially reflected in each of our particular consciousnesses? [. . .] Union, the true upward union in the spirit [consciousness], ends by establishing the elements it dominates in their own perfection. *Union differentiates.*[35]

But not only are each of the centers of consciousness preserved in their union, they are *evolutionarily enhanced*; the *n* centers join, but in that joining, although they retain their own personalities, an additional personality, *n* + 1, is formed, and "*since there is no fusion or dissolution* of the elementary personalities, the centre in which they join *must necessarily be distinct from them, that is to say have its own personality.*"[36]

It is at this point in his essay that Teilhard introduces the term *Omega*, describing it as the cosmic point of total synthesis without which "the world would not function," and describing its relationship to the noosphere:

---

*Note that it is here, in Teilhard's earliest known essay, written at Dunkirk in April 1916, that Teilhard first speaks at length about the centers and the sphere: "We are the countless centres of one and the same sphere [. . .] The totality of a sphere is just as present in its centre, which takes the form of a point, as spread over its whole surface."[34]

The noosphere in fact *physically* requires, for its maintenance and functioning [. . .] the unifying influence of a *distinct* centre of super-personality [. . .] a centre different from all the other centres which it "super-centres" by assimilation: a personality distinct from all the personalities it perfects by uniting with them [. . .] Consideration of this Omega will allow us to define more completely [. . .] the hidden nature of what we have till now called, vaguely enough, "human energy."[37]

The Omega Point

Central to the architecture of Teilhard de Chardin's hyperphysics is the concept of Omega or the Omega point. According to his close friend Henri de Lubac, Teilhard's first use of the term can be found in one of his earliest essays:

> In the first essay that was entitled *Mon Univers* (1916) he carefully distinguishes, in order to study their relationships, "Omicron, the natural term of human and cosmic progress," from "Omega, the supernatural term of the Kingdom of God" or "Plenitude of Christ." Later, he was to abandon this particular terminology, but he retained the distinction it expressed.[38]

Twenty years later Teilhard was using the term in a more secular, scientific context, as seen here at the close of a lecture delivered at the French embassy in Peking on March 10, 1945.[39] "Ahead of, or rather in the heart of, a universe prolonged along its axis of complexity, there exists a [. . .] centre of convergence [. . .] let us call it the *point Omega*."[40]

Teilhard devotes an entire section to "The Attributes of the Omega Point," in his book, *The Human Phenomenon* begun in Paris in 1939 and completed in China during World War II. In a section that in his first draft had been called, "Spirit and Entropy," Teilhard says of Omega:

> Expressed in terms of internal energy, the cosmic function of Omega consists in initiating and maintaining the unanimity of the world's reflective particles under its radiation. But how could it carry out this action if it were not somehow already [. . .] *right here and now*? [. . .]
>
> Autonomy, actuality, irreversibility and finally, transcendence are

the four attributes of Omega [. . .] Omega is the principle we needed to explain both the steady advance of things toward more consciousness and the paradoxical solidity of what is most fragile [. . .] Something in the cosmos, therefore, escapes entropy—and does so more and more.[41]

An important element of Teilhard's model of human energy lies in his understanding that consciousness can be expressed in thermodynamic terms. In the essay "Human Energy," in a section with the title, "The Maintenance of Human Energy and 'The Cosmic Point Omega,'" Teilhard describes how an axial form of this heat energy powers the current of conscious human energy. Teilhard describes the generation of this current:

Considered in its organic material zones, human energy obeys the laws of physics and draws quite naturally on the reserves of heat available in nature. But studied in its axial, spiritualized form, it is found to be fed by a particular current (of which thermodynamics might well be, after all, no more than a statistical echo), which, for want of a better name, we will call "tension of consciousness."[42]

Teilhard here once again links thermodynamics to the phenomenon of consciousness, and he goes on to refute the widespread scientific paradigm of consciousness as a mere epiphenomenon of the material universe:

We still persist in regarding the physical as constituting the "true" phenomenon in the universe, and the psychic as a sort of epiphenomenon [. . .] we should consider the whole of thermodynamics as an unstable and ephemeral by-effect of the concentration on itself of what we call "consciousness."[43]

It is reasonable to consider Teilhard's Omega point, resolved down to the smallest possible dimension known to physics, as equivalent to a "Planck holosphere," discussed in chapter 6.

In one of his last essays, "The Nature of the Point Omega," Teilhard states that it is in the noosphere that all is truly preserved, for it is here that all experience is gathered and saved eternally:

In convergent cosmogenesis, as I have said, everything happens as if the preservable contents of the world were gathered and consolidated, by evolution, at the centre of the sphere representing the universe [. . .] a cosmic convergence [. . .] to bind objectively to the real and already existing centre.[44]

Here again can be seen a congruence between Teilhard's process viewed as a convergent cosmogenesis and Bohm's process seen as an enfolding of the explicate into an implicate domain.

Teilhard concludes the essay "Human Energy" with a highly optimistic observation revealing once again his lifelong fascination with the concept of the human personality's mode of survival beyond biological death:

The principle of the conservation of personality signifies that each individual nucleus of personality, once formed, is forever constituted as "itself"; so that, in the supreme personality that is the crown of the universe, all elementary personalities that have appeared in the course of evolution must be present again in a distinct (though superpersonalized) state [. . .] each elementary person contains something *unique and untransmittable* in his essence.[45]

"Centrology: Dialectics of Union" (1944)

During the occupation of China by the Japanese, Teilhard's anthropological work was severely curtailed, and he found himself with time to go deeply into the development of his more abstract ideas, which he expressed systematically and in great detail in his 1944 essay, "Centrology," written under somewhat stark wartime conditions during his Peking confinement. At the beginning of his essay, he boldly states what he considers to be the scientific nature of this essay: "It is not an abstract metaphysics, but a realist ultraphysics of union."[46]

Teilhard opens his essay with an immediate discussion of "Centres and Centro-Complexity," describing how in living elements of the biosphere we find a continuation of the "granular (atomic molecular)" structure of the universe, and that, in fact, the human body "is simply a 'super-molecule.'"[47] However, unlike conventional physicists, who see cosmic particles as sources or centers of radiation and then map that radiation in the spacetime domain, Teilhard places the focus of his inquiry on the "within" of each so-called particle. According to Teilhard, the spacetime particles are not only centers of origination of radiation but each one of them also "has" a within, and "is" a within, a within that is a mode of consciousness, a psychic centre: "They are psychic centres, infinitesimal psychic centres of the universe [. . .] in other words, consciousness is a universal molecular property."[48]

Teilhard goes further to claim that an increase in consciousness can be found associated with an increase in "centro-complexity," and he defines "the coefficient of centro-complexity" as "the true absolute measure of being in the beings that surround us."[49] Teilhard describes biology as "simply the physics of very large complexes."[50]

He points out that the atomic complexity of a virus is of the order of 10^5 atomic particles, and this complexity increases dramatically by the time we reach the size of a cell at 10^{10}, but in the brains of large mammals, has reached the great complexity of 10^{20} particles.[51]

Teilhard states that "if the universe is observed in its true and essential movement through time, it represents a system which is in a process of internal centro-complexification," and asserts a definition of evolution to be "a transition from a lower to a higher state of centro-complexity."[52]

In his essay, "Man's Place in the Universe," Teilhard had argued that existence entails three infinities: the infinite large, the infinite small, and the infinite complex, and he illustrates this with the chart reproduced in figure 8.3 below.[53] Using data from objects in nature Teilhard plots a curve with two axes: a vertical y-axis scaled in Size (length in centimeters), and a horizontal x-axis on which is measured increases in Complexity (total number of atoms per object). Note that both scales are calibrated in base 10 logarithms. The curve plots size versus complexity for various natural entities, named on both axes.

Points *a* and *b* on the plot indicate where Teilhard believes "state changes of consciousness" have occurred. Point *a* marks the emergence of life, and point *b* indicates the emergence of reflective consciousness (i.e., thought, being able to think about thinking).

184 • Hyperphysics of the Noosphere

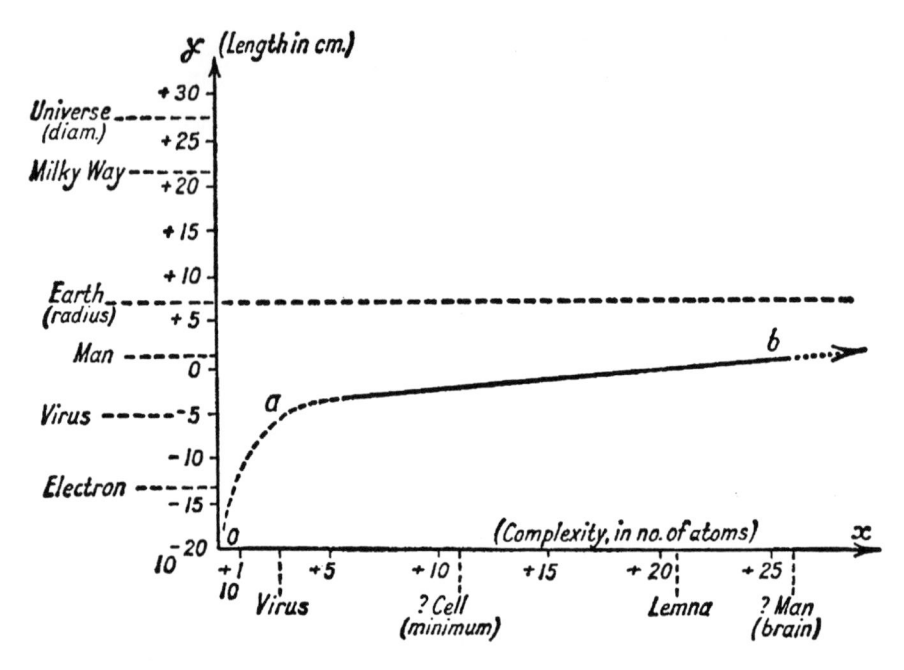

Fig. 8.3. Teilhard's natural curve of complexities. Image from Teilhard, "Man's Place in the Universe," 226.

The curve in figure 8.3 can be extrapolated, leading to a planetary change of state, a state described here by Teilhard as an awakening of a consciousness common to the whole earth:

> We can see it only as a *state of unanimity*: such a state, however, that in it each grain of thought, now taken to the extreme limit of its individual consciousness, will simply be the incommunicable, partial, elementary expression of a total consciousness which is common to the whole earth, and specific to the earth: *a spirit* [consciousness] *of the earth*[54]

Elsewhere Teilhard has described the importance of the concept of reflection (alternately translated into English as reflexion), and this is associated with point *b* on the complexity chart, the concept also applies to a projected point *c* that would be a change of state for the consciousness of the planet, a noospheric "reflection":

> "Reflection," as the word itself indicates, is the power acquired by a

consciousness of turning in on itself and taking possession of itself *as an object* endowed with its own particular consistency and value: no longer only to know something but to know *itself*; no longer only to know, but to know that it knows.[55]

Teilhard's chart thus supports the first general conclusion in his essay on "Centrology," that the universe is in a state of internal "centro-complexification," and that in this "transition from a lower to a higher state of centro-complexity," we see a concomitant increase in complexity-consciousness, in a process that Teilhard terms cosmogenesis through centrogenesis.[56]

From his earliest essays, Teilhard sees the planet itself as an evolving, larger entity, out of which humanity has sprung and to which humanity is adding new capabilities. Teilhard presents geological and zoological evidence of the planet Earth as an evolving life-form, a global being in the transitional process of awakening into a planetary state of reflective self-consciousness.[57] He identifies a distinct axis of successive forms, layers of increasing complexity and centrification running from geogenesis through biogenesis and beyond into psychogenesis; this axis, he insists, can be seen continuing in the present noetic awakening that it is life itself that is engendering the birth of a new mode of planetary consciousness, comprising an entity that he himself has named the noosphere.[58] He speculates that eventually "life might use its ingenuity to force the gates of its terrestrial prison [. . .] by establishing a connection psyche to psyche with other focal points of consciousness across space," and notes "the possibility of 'centre-to-centre' contacts between perfect centres."[59]

But it is not only the planet itself that is evolving for Teilhard, he also views consciousness in humanity as evolving and thus sees the human species accelerating toward an evolutionary threshold, where it will experience the nature of energy and self-reflection in ever newer ways, while feeling itself drawn magnetically toward new states of greater cohesion and complexity, not only of radiant physical energy, which he terms "the tangential component," but of an increasingly conscious, psychic flow, the spiritual or "radial component" of energy.[60] He even senses an imminent transformation in the biophysical gateway, the human brain, in which he foresees form and function itself complexifying past the point of isolated self-reflection:

Is there not in fact, beyond the isolated brain, a still higher possible complex: by that I mean a sort of "brain" of associated brains? [. . .] From this point of view, the natural evolution of the biosphere is not only continued in what I have called Noosphere, but assumes in it a strictly convergent form which, towards its peak, produces a point of maturation (or of collective reflection).[61]

Another term used by Teilhard in describing this process is "convergence." He states that, "In the organo-psychic field of centro-complexity, the world is convergent; the isospheres are simply a system of waves which as time goes on (and it is they which measure time) close up around Omega point"; the world, according to Teilhard, is moving continuously in "a transition from a lower to a higher state of centro-complexity."[62]

Centrogenesis

Teilhard coins a new term, *centrogenesis*, to encapsulate this process. In "Centrology," Teilhard begins his discussion of centrogenesis by claiming that the universe is made up of psychic nuclei, similar to the theory of monads developed by Gottfried Wilhelm Leibniz; monads were described by Leibniz as being the most basic, fundamental entities of which the cosmos is constructed (an idea seen here as a precursor to the model of holospheres in the holoplenum), however, in his theory the monads are completely independent of one another, though in complete harmony.[63] Unlike the monads of Leibniz, Teilhard's nuclei are interconnected in three simultaneous ways.[64] These relationships are as follows:

- tangentially—"on the surface of an isosphere";
- radially—"through nuclei of lower centro-complexity" ($n1$, $n2$, etc.);
- radially—"inwardly, creating an isosphere of a higher order" ($n + 1$).

Teilhard describes these elementary cosmic centers as "partially themselves," and "partially the same thing." Figure 8.4 shows four drawings presented by Teilhard at the beginning of his essay on "Centrology."

Three of Teilhard's images shown in figure 8.4 (Teilhard's figures 2, 3, and 4) depict the three stages of centrogenesis. Here we can see the

Figure 1. Diagram symbolizing the principle phases of centrogenesis (convergence of the universe along its axis of centro-complexity or personalization).

Note the concentric system of isospheres (surfaces of equal centro-complexity), subdivided into three zones by the two critical surfaces of centration and reflection (cf. sections 9 and 13).

Omega point is at the centre.

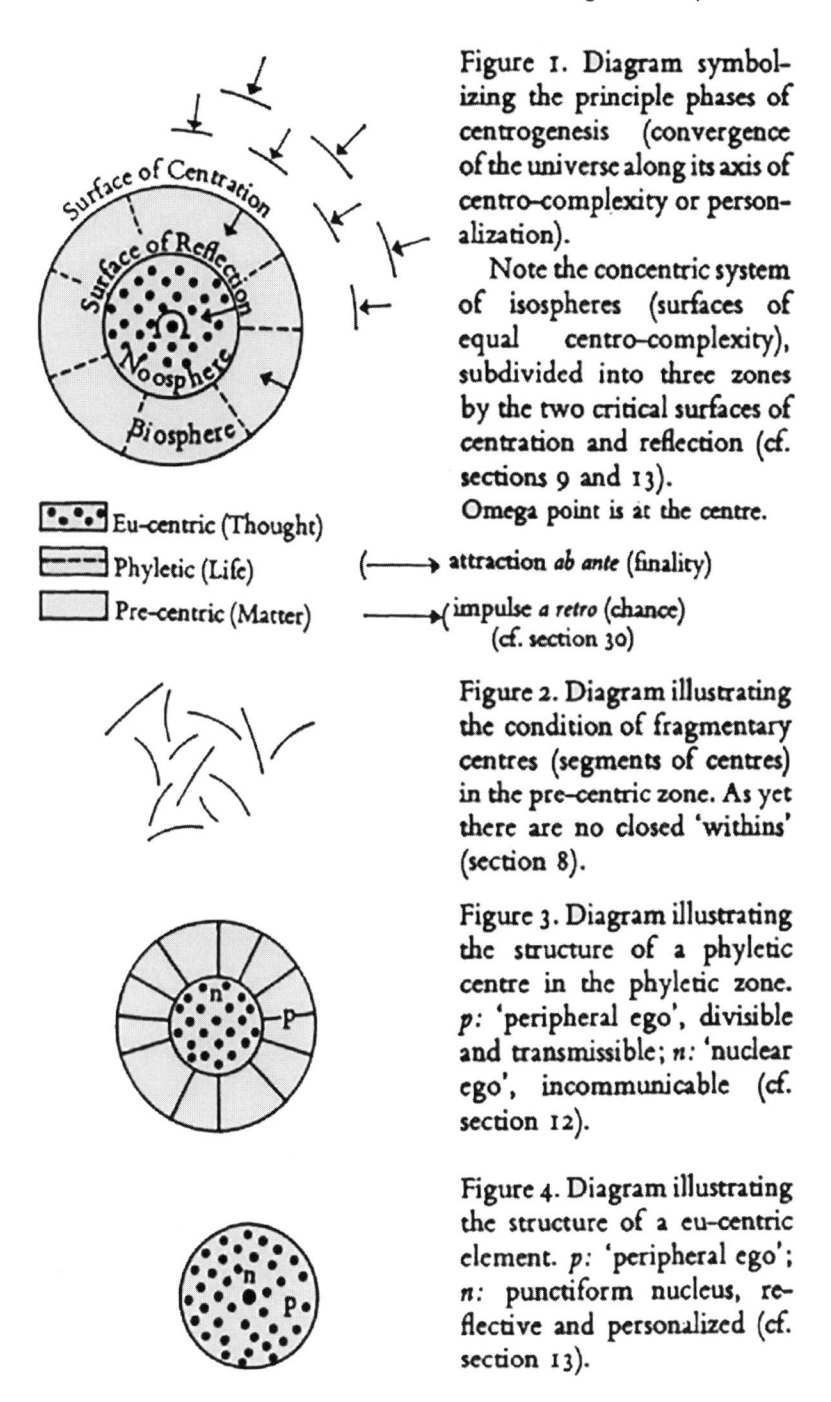

(⟶ attraction *ab ante* (finality)

⟶(impulse *a retro* (chance)
(cf. section 30)

Figure 2. Diagram illustrating the condition of fragmentary centres (segments of centres) in the pre-centric zone. As yet there are no closed 'withins' (section 8).

Figure 3. Diagram illustrating the structure of a phyletic centre in the phyletic zone. *p*: 'peripheral ego', divisible and transmissible; *n*: 'nuclear ego', incommunicable (cf. section 12).

Figure 4. Diagram illustrating the structure of a eu-centric element. *p*: 'peripheral ego'; *n*: punctiform nucleus, reflective and personalized (cf. section 13).

Fig. 8.4. Teilhard's figures 1 through 4 in "Centrology."
Image from Teilhard, "Centrology," 100.

condition of "fragmentary centres (segments of centres)," not yet enclosed in isospheric configurations, and still devoid of what Teilhard calls "personality." These are elements that he terms "pre-centric" fragments, having no "withins."[65]

More evolved is the second image from the bottom, which reveals *phyletic centricity*, a change of state brought about by the self-closing of numerous fragmentary centers, which he defines as "life," and which manifests as phylum.[66] In regarding this state, Teilhard brings up two questions: (1) He asks, "How can we conceive the passage and communication of a 'within' from mother-cell to daughter-cell?," and (2) "Under what conditions is the phylum provided the greatest possible richness and variety for the evolutionary transmission of successful properties?"[67]

To answer the first question, Teilhard observes that there are "two sorts of ego in each phyletic centre, a nuclear ego [. . .] and a peripheral ego."[68] The *peripheral ego* is incompletely individualized, separate, and according to Teilhard, it is therefore divisible and can be shared through replication or association. Teilhard then explains how the second ego, the *nuclear ego*, communicates: "It is in virtue of the arrival at zero of its centric diameter that the living centre, in its turn, attains the condition and dignity of a 'grain of thought.'"[69]

Thus the particular phyletic center (consisting of a peripheral ego and a nuclear ego) retains access to all of the information ever associated with the particular phylum through resonance among phyletic isospheres (nuclear egos).

This is in agreement with both Sheldrake's theory of morphic resonance and Bohm's quantum cosmology, as it provides a mechanism whereby speciation information may be shared through resonance, transferred into the explicate domain via the implicate domain, and vice versa.[70] Since the implicate order of the nuclear ego is nondual (outside of the spacetime domain), it has random access to information generated in *all time* and *all space* and is thereby able to apply total information in its processing. Reflective consciousness is also a characteristic of this phyletic center, and the typical human personality can be here identified with Teilhard's "peripheral ego," while Omega provides a guiding force via centrogenesis (Bohm's quantum potential function, Q, discussed in the next chapter).

To answer his second question, (how does the phylum provide the greatest possible richness and variety for evolutionary growth?), Teilhard identifies a "twofold complexity," one spatial, and the other temporal. Spatial complexity refers to the spread of the phylum over the surface of its isosphere, the creation of a population of phyletic centers that gather experience and mutate in the ever-changing environment. At the same time, the action of temporal complexity provides the vast number of "trials" over the myriad of generations of which the phylum's ancestors represent the total sum.[71]

At the bottom of Teilhard's figure 8.4 we see the "structure of a eucentric element," as a major, and perhaps ultimate, "change of state" in the emerging process, a reflection of consciousness in the noosphere, a reflective connection with Omega.[72] In the typical human grain of thought, "reflection" has not yet reached a resonance with Omega. However it is possible to effect, in the individual, as Teilhard says, a "eu-centered, 'point-like' focus [. . .] and this is enough to allow the appearance of a series of new phenomena in the later advances of centrogenesis."[73] It is not difficult to assume that the "point-like focus" recommended here by Teilhard is a reference to his own direct experience in effectuating such a focus, his own participatory experience of consciousness.

Teilhard laments that while humans are generally reflective, only a few have yet been able to integrally connect with the punctiform nucleus, Omega; but those able to connect for any duration find that they now "possess the sense of irreversibility," a conviction that makes "an escape from total death [. . .] possible for a personalized being."[74] At this point, says Teilhard, "Welded together in this way, the noosphere *taken as a whole*, begins to behave tangentially, like a single megacentre [. . .] ontogenesis of collective consciousness and human memory."[75] Here, Teilhard makes a prediction, stating that the evolution of human society on the planet will eventually lead to the following:

> The accelerated impetus of an earth in which preoccupation with production for the sake of well-being will have given way to the passion for the discovery for the sake of fuller being—the super-personalization of a super-humanity that has become super-conscious of itself in Omega.[76]

In a psychically convergent universe, the process of a reflective connection of the peripheral, phyletic ego with the central, eu-centric ego leads ultimately to a "final concentration upon itself of the noosphere," Omega. "Omega appears to us fundamentally as the centre which is defined by the final concentration upon itself of the noosphere—and indirectly, of all the isospheres that precede it."[77]

Here Teilhard writes of "all the isospheres that precede it," which, of course, are all of the isospheres in time and space that are ourselves, our ancestors, and other centers of phyletic centro-complexity. When a locus of fragmentary centers closes, joining together to form a phyletic center, the newly formed noosphere experiences a state change and awakens. From that point, moving forward in space and time, the phyletic noosphere evolves through a series of internal noospheres, growing higher in energy and complexity, until it reaches its ultimate "final concentration" at Omega, as Teilhard states, "In Omega then, a *maximum complexity*, cosmic in extent, coincides with a *maximum cosmic centricity*."[78] In other words, at the heart of matter, at the Omega point, Teilhard tells us that maximum complexity equals maximum centricity.

Teilhard refers to the center as "a quantum of consciousness," and tells us that each degree of consciousness at a given moment only exists as "an introduction to a higher consciousness."[79] He says that this general process is irresistible and irreversible.[80]

> The initial quantum of consciousness contained in our terrestrial world is not formed merely of an aggregate of particles caught fortuitously in the same net. It represents a correlated mass of infinitesimal centres structurally bound together by the conditions of their origin and development.[81]

Moving forward in his essay on centrology, Teilhard goes on to describe four attributes of Omega as:

- personal
- individual
- already partially actual (spacetime energy)
- partially transcendent (nondual)[82]

It is personal "since it is centricity that makes beings personal," and "Omega is supremely centred."[83] It is individual because it is "distinct from (which does not mean cut off from) *the lower personal centres* which it super-centres."[84] These lower centers are the various phyletic, peripheral egos, each of which is uniquely individual, yet can join with Omega without losing their individuated personality; in fact it is in the relationship, the resonance with Omega, that the very uniqueness of the individual is highlighted (i.e., "union differentiates").

Omega is both "partially actual" and "partially transcendent." The relationship is one Teilhard characterizes as "a 'bi-polar' union" of the emerged and the emergent. It is partially transcendent beyond the very center of spacetime, within the Bohmian implicate order. There, all is "partially transcendent of the evolution that culminates in it." And it is there that all spacetime experience is gathered, in the partially transcendent, in continuous communication via a mathematically dynamic process. Otherwise, Teilhard tell us, there would not "be the basis for the hopes of irreversibility."[85]

And it is Omega that provides the momentum for centrogenesis.

> Drawn by its magnetism and formed in its image, the elementary cosmic centres are constituted and grow deeper in the matrix of their complexity. Moreover, gathered up by Omega, these same centres enter into immortality from the very moment when they become eu-centric and so structurally capable of entering into contact, centre to centre.[86]

Centrology and Complexity: Being and Union

At this point in his essay, Teilhard inserts what he calls "A Note on the 'Formal Effect' of Complexity" in which he examines the underlying roots of his assertion that consciousness *increases* with complexity in union (centro-complexity), and he states that an understanding of this phenomenon has come to him "experientially."[87] Teilhard sets forth, in two Latin propositions, the fundamental ontological relationship between being and union:

1. The one passive: *Plus esse est a* (or *ex*) *pluribus uniri.*
2. The other active: *Plus esse est plus plura unire.*

Plus esse can be translated as "more being," "growth in being," or "being increases," but from the context of this essay the phrase *plus esse* might be translated as "consciousness increases."[88] Thus Teilhard describes a bi-modal process of an increasing consciousness in the universe: actively and passively.

In the first proposition, the verb *uniri* is in the passive voice, "be united," and in context can be translated, "become one, become a center." The *a/ex* prepositions, often used interchangeably in Latin, indicate "out of" and "from." Thus, *a pluribus* can be translated "out of many, from many," and accordingly, the first proposition may be translated as, "A new conscious center grows by many being joined."

In the second proposition, however, the verb *unire* is in the active voice (i.e., "unite"), which in context can be translated "make a centered unity" of *plura* (literally, "more/many things"). Thus, the second proposition can be translated as "A new consciousness center grows by many uniting." Teilhard next applies these two propositions to the following stages of centro-complex evolution:

1. The appearance of life through association of fragments of centers.
2. The deepening of phyletic centers.
3. The emergence of reflective centers.

In the first instance, for a state change occurring in the domain of prelife, Teilhard again formulates a metaphysical axiom in Latin: *Centrum ex elementis centri*, which translates as, "The center out of elements of the center." In this domain centers are "built up additively, through the fitting together and gradual fusing of 'segments' of centres."[89] This is a passive growth.

In the second stage, Teilhard says, "being born from an egg (*centrum a centro*) complexifies upon itself by cellular multiplication." Here, each center complexifies itself by increasing its own depth of complexity.[90] Here the active growth emerges, a growth in part directed by the center. A similar pattern can be seen in Bohm's quantum potential function, a guiding energy from within the implicate order, within Teilhard's Omega point.

In the third stage it is in what he calls the *eu-centric* that this quantum potential metaphysical process becomes super-active, and as Teilhard says, it is from "the noospheric centre, Omega," that there emerges "*Centrum super centra*," translated as "a new center emerges from an old center."[91] Teilhard is saying here that Omega is not simply the sum of components, but something *new*, a unique entity bursting forth:

> In the eu-centric domain, the noospheric centre Omega, is not born from the confluence of human "egos," but emerges from their organic totality, like a spark that leaps the gap between the transcendent side of Omega and the "point" of a perfectly centered universe.[92]

Teilhard describes how, for individual human phyletic "egos," it may be possible to go beyond the present general evolutionary stage of consciousness, developed through the general societal drift of hominization through time. Teilhard explains how such an evolutionary leap might be accomplished:

> This can be envisaged in two ways, either by *connecting up neurones* that are already ready to function but have not yet been brought into service (as though held in reserve), in certain already located areas of the brain, where it is simply a matter of arousing them to activity; or, who can say?, by direct (mechanical, chemical or biological) stimulation of new arrangements.[93]

In his phrase, "direct stimulation," we can imagine a range of approaches that might be used to catalyze the evolutionary growth and transformation of consciousness within an individual, who would then experience what Teilhard terms an "ultra-hominization" of reflective consciousness, evolutionarily beyond that of the currently conventional ranges of human experience.

His first suggestion, "connecting up neurones that are already ready to function," might be seen as a catalysis by birth (genetic predisposition), accidental circumstances (serendipitous encounters with the sublime), or specific psychophysical techniques (e.g., prayer, yoga, physical exercises, special diets, fasting, sweat lodges, etc.); but Teilhard goes even further to suggest that the evolutionary process might be boosted within the individual human personality "by direct (mechanical, chemical, or biological) stimulation," and here we are reminded of exploration and critical experimentation with psychotropic drugs (psilocybin mushrooms, LSD, cannabis, ayahuasca, etc.), or through direct energy-stimulation devices as can be seen in the recent technologies of transcranial magnetic stimulation, or transcranial direct-current stimulation.[94]

Similarly, but moving from neuronal to human level, Teilhard forecast's the connecting up of a network of individual consciousness via "a direct tuning": "In nascent super-humanity [. . .] the thousands of millions of single-minded individuals function in a nuclear way, by a direct tuning and resonance of their consciousness."[95]

At the close of "Centrology," Teilhard sets forth his "Corollaries and Conclusions." He begins by summarizing in a sequence the stages depicted earlier in figure 8.4. Here he focuses on the evolution of consciousness leading to the birth of reflective thought in mankind and the dawn of what he calls "Omega." These five points on the arc of centro-complexity are seen as categorically distinct evolutionary changes of state:

1. The appearance of life through association of fragments of centers.
2. The deepening of phyletic centers.
3. The emergence of reflective centers.
4. The birth of mankind (and reflective thought).
5. The dawn of Omega.[96]

In each of these five steps can be seen the effect of an increase in union, the creative energy of union causing changes of state, not simply due to some rearrangement or summation of parts, but, as Teilhard says, "doing so under the influence of the radiation of Omega."[97] The holoflux theory analog of Teilhard's "radiation," of course, is Bohm's quantum potential function, "Q," radiating from the implicate order to emerge within spacetime. Teilhard's "Omega point" can thus be viewed as a spherical portal, an analog of Bohm's "implicate order."[98]

In discussing the transition from stage 4 to stage 5, the change of state from the "birth of thought" to the "dawn of Omega," Teilhard insists that there should be *no* concern that such a change of state would mean a loss of personality, or a death of our uniquely distinct egos; on the contrary, says Teilhard, the noosphere is comprised of an effective *re-union* of all individual "savable elements" of each personality. Even more, it serves to effect an even higher, more complete integration of *each* individual experience, a heightening of individual personality, "a cosmic personalization, the fruit of centrogenesis."[99] It is, stresses Teilhard, through each sub personality center-to-center contact *within* the noosphere (through the center), that each individual personality is "super-personalized."[100] This center-to-center contact, Teilhard tells us, is a perceptual condition of each individual center merging into the noosphere and it *is not an epiphenomenon* of such a union, but that it is a *requirement* for reaching full integral personalization:

> Something (someone) exists, in which each element gradually finds, by reunion with the whole, the completion of all the savable elements that have been formed in its individuality; thus the interior equilibrium of what we have called the Noosphere requires the presence *perceived by individuals* of a higher polar centre that directs, sustains and assembles the whole sheaf of our efforts.[101]

This process can be seen as a cybernetic feedback loop: the "higher polar centre" receives input from all of the "savable elements" of each "individuality," and, working with this information, the higher personality ("polar center") "sustains and assembles the whole sheaf of our efforts." In this way evolution proceeds through a continuous cyclic process of individual centers developing in spacetime, merging into their respective centers through the directional "drift" of centro-complexity, *but not being lost* in the merger. "To the extent that the grain of consciousness is *personalized*, it becomes released from its material support in the phylum [. . .] detached from the matrix of complexity, and meets the ultimate pole of all convergence."[102]

Finally, the fifth stage in the process of centro-complexity, "the dawn of Omega," occurs at the point where *thought* transforms into omni-contact with all other centers, as well as with the "higher" $(n + 1)$ center of personality, at which point there is a flaring forth into a categorically new mode of reflective consciousness, effecting a major change in state.

Teilhard concludes that, "Far from tending to be confused together, the reflective centres intensify their *ego* the more they concentrate together."[103]

Isospheres

A key concept developed in "Centrology" is Teilhard's model of "isospheres," which he defines as "surfaces of equal centro-complexity."[104] He sees evolution catalyzed on these isospheres when "a maximum density of particles with a corresponding maximum of tentative gropings is produced on each isosphere."

These various regions of the planet each have their own unique and identifiable physical characteristics (temperature, density, etc.), and each might be considered as an isosphere of the planet. Teilhard's use of the word *isosphere* is focused less upon the physics of geology, and more upon the physics of metaphysics, a topology of consciousness, as he states in "Centrology": "Thus there emerges the pattern of a *centred universe*—elements of equal complexity (and hence of equal centricity) being spread out over what we may call isospheres of consciousness."[105]

Here Teilhard has clearly indicated the direction in which he sees increased consciousness: toward the center, in the direction of Omega. According to quantum theory, the radial distance between any two nested isospheres may be no less than the Planck length (the smallest possible dimension of space, and this leads us to visualize an enormous yet finite series of isospheres in spacetime, nested like Russian dolls, or perhaps like separate capacitor plates in an electronic circuit, beginning at the boundary of the Planck holosphere (enclosing Omega, the implicate order) and reaching an outer limit at the current (continuously expanding) diameter of the universe.

In this cosmological topology of nested three-dimensional isospheres, electrons would not travel in planar, two-dimensional, circular orbits, as depicted in Bohr's model, but are "smeared out" as holonomic patterns of tangential flux over the surface boundaries of three-dimensional shells covering the surface areas of the isosphere.[106]

As we move outward from the central Planck holosphere into isospheres of higher radial dimensions, each holosphere must be separated, minimally, by one Planck length. The manifestation of these isospheres in spacetime would provide the geometric capacity for storing multiple qubits of information, as previously developed by John Wheeler in considering the event horizon of a black hole. Thus, we may envision quantum states in a series of holospheric shells extending from the central Planck holosophere to the current boundary of the universe as spheres rather than rings. The universe here can be seen to consist of an almost infinite holoplenum of intersecting concentric shells of implicate order holospheres throughout spacetime.

As seen in Bohr's model, to move from one shell to another shell, the electron cannot move *continuously* through some intervening spacetime gap, but instead is seen to execute a quantum leap to the adjacent level, transitioning from one isospheric shell to another during a single Planck time cycle at the rate of 5×10^{-44} seconds. The energy formations appear at the next holosphere shell at the "clock-speed" of light, yet not moving linearly within spacetime as normally understood, but moving alternately between explicate spacetime and implicate nondual domains of being, nudging every center toward an ever so slightly new direction in their evolutionary arcs in spacetime.

Such a transformation might also be identified with Jean Gebser's "mutation of consciousness," the evolutionary mutation into what Gebser calls "integral consciousness."[107] In full agreement with Teilhard's assertion that evolution precludes loss of experienced personality, Gebser assures us that in evolving structures of consciousness, previous properties and potentialities do indeed survive: "In contrast to biological mutations, these mutations of consciousness do not assume or require the disappearance of previous potentialities and properties, which, in this case, are immediately integrated into the new structure."[108]

Like Gebser, Teilhard observes that all personalities are incomplete, that they are continually evolving, however slowly, and compares our individual personalities to the fragments of centers that seek for one another in the pre-living zones of matter, reminding us that, "at our level of evolution we are still no more than rough drafts."[109] Teilhard here states with emphasis, *"the personal—considered quantitatively no less than qualitatively—is continually on the upgrade in the universe."*[110]

Breaking Teilhard's "Death-Barrier"

One of Teilhard de Chardin's mature conclusions is that our entire consciousness is *not* snuffed out when our material bodies die. If consciousness manifests as energy, then following Einstein's observation that "energy cannot be created or destroyed, it can only be changed from one form to another," any signal composed of an energy of consciousness must follow the same pattern.[111] While there may be a transformation of energy, there can be no absolute destruction, because energy (specifically the modulated energy flux of a conscious entity) cannot suddenly vanish. Such a view is, simply, superstition.

By the very logic of evolution, in order for the species to learn, adapt, and preserve experiences gathered in the spacetime domain, evolution at the human stage must break the "Death-Barrier," a term Teilhard develops in one of his final essays, "The Death-Barrier and Co-Reflection," completed January 1, 1955, three months before his own death on Easter Sunday, April 10, 1955: "When biological evolution has reached its *reflective* stage ("self-evolution") it can continue to function only in so far as man comes to realize that there is some *prima facie* evidence that the death-barrier *can* be broken."[112]

Teilhard had written previously of his own participatory experiences supporting his belief in an immortality of consciousness in a letter to his paleontologist colleague and friend, Helmut de Terra:

> My visible actions and influence count for very little beside my secret self. My real treasure is, *par excellence*, that part of my being which the centre, where all the sublimated wealth of the universe converges, cannot allow to escape. The reality, which is the culminating point of the universe, can only develop in partnership with ourselves by keeping us within the supreme personality: we cannot help finding ourselves personally immortal.[113]

Teilhard assumes that if there is a part, or region, or mode, or domain of our consciousness that continues beyond our bodies, beyond the death of our bodies, as he says in "Breaking the Death-Barrier," then should we not then be motivated to know and even explore that domain? That is the real treasure he is sharing with us: once specific memories are gone, the personality lives on, "keeping us within the supreme personality," within the implicate order, at the center, everywhere. But this "personality of the transcendent" can be seen, by mystics at least, even before the approach of the Death-Barrier.[114]

Not only does Teilhard categorically reject the majority of contemporary humankind's tacit assumption that death is "the end" (i.e., the end of individual consciousness); he worries that such an erroneous stance might delay what he saw as the natural emergence of a noosphere cultivated and powered by human conscious energy, a collective holonomic energy modulated by and powered by living *Homo sapiens*.

9

The Superposition of Consciousness

The Superposition Principle

It is commonly understood that all awareness is consciousness of something—be it a sound, an image, a sensation, an emotion, a dream, or an internal verbal thought. These experiences often overlap, occurring in what seems to be the same perceptual moment. Although each stream of experience remains distinct, they are somehow integrated into a cohesive whole. A clear example is listening attentively to music and being able to discern several distinct instruments while simultaneously appreciating the dynamics of their interplay.

What actually is it that is both observing and distinguishing between the separate tracks of these simultaneously occurring streams of experiences? And what is it that is able to compare the present stream in time with an immediately earlier stream in time, and thus discern changes with which to make comparisons? It appears that there is a fundamental form of consciousness at work, one that surpasses the mere combination of individual experience streams. This meta-consciousness, distinct from the individual streams, somehow gains a broad, panoramic view of all the input streams. Additionally, this primary consciousness possesses the remarkable ability to focus on and fine-tune one or more of these streams by dampening and filtering out the others. Clearly, this higher level of consciousness has evolved multitasking abilities—skills that can be further refined and strengthened through practices like insight meditation, yoga, and similar disciplines.

The Superposition of Consciousness • 201

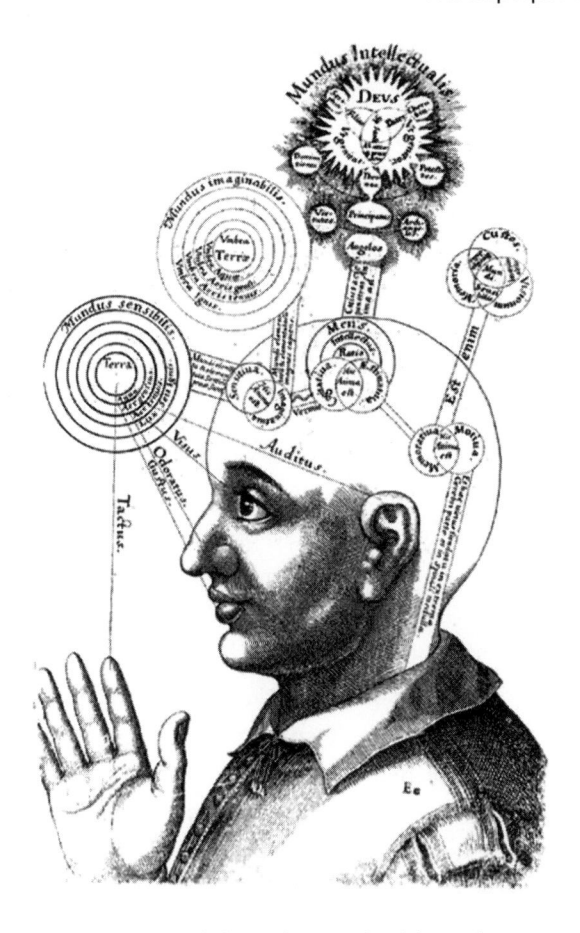

Fig. 9.1. Superimposed simultaneous streams of thought.

Streams of thought can be likened to streams of data flowing through the internet. Just as a fiber-optic cable can transmit millions of documents in a two-way flow of information, a variety of information-rich signals (such as video, audio, web pages, and raw data) travel simultaneously through the same cable without distorting one another. This capacity to carry a vast array of different information streams at the same time, without distortion or loss, is known as the *superposition principle*.

The superposition principle functions specifically in the frequency domain. To apply this principle, incoming data streams must first be converted from their original time-domain signals into frequency signals using the Fourier transform. Once converted, these frequency signals are combined and transmitted through the communication link. This approach overcomes the challenge of combining time-domain signals, which is complex due to the non-linear nature of spacetime—where straight lines do not exist and everything is curved. As a result, the mathematics involved in combining time-domain signals is highly intricate, requiring advanced calculus and differential equations.

However, when time-based signals are transformed into discrete frequencies, the process of combining them becomes much simpler. Instead of dealing with complex calculus, the frequencies can be easily combined through simple addition. In other words, the signals, once converted to their frequency domain versions via the Fourier transform, can now be straightforwardly added together. This is the superposition principle, the bedrock of our device powered societies, and it is the Fourier transform that is the mathematical bridge that is used to transform the time-based signals from out of time and into the frequency domain.

The Fourier transform, a groundbreaking mathematical equation, was first discovered and formalized by the French mathematician Jean-Baptiste Fourier. This remarkable development originated from Fourier's work in 1807, when he sought a method to reliably cast large cannons for Napoleon's army. Fourier devised a technique to accurately predict the amount of heat and time required at a specific furnace temperature to produce strong, reliable cannons. His mathematics solved a major problem that had been plaguing the production of these large cannons.

Cannon technology had been evolving for nearly a century, influenced by advancements in Japanese samurai metallurgy, which spurred European armies to experiment with various metallic additives, including carbon, in order to forge increasingly larger and more powerful iron and bronze cannons. However, casting mixed metals together posed significant risks. If the temperature was too low, the metals wouldn't melt uniformly, leading to cracks in the cannon barrel, rendering it useless. Conversely, if the temperature was too high, the metal weakened, causing some barrels to explode when heated during firing—a dangerous outcome for the cannon crews.

Fourier's solution was to create a mathematical equation that could determine the optimal temperature and timing needed to cast a cannon barrel from a mix of metals, ensuring that the metals fused into a seamless alloy. By applying Fourier's equation, foundries could produce cannon barrels with greater inherent strength, free of cracks, and—importantly for the safety of the operators—resistant to exploding during use.

The challenge of mixing metals arises because each metal has its own distinct phase diagram. A phase diagram illustrates how a metal's behavior changes with variations in temperature and pressure, graphically showing the different phases a metal undergoes as it cools (see fig. 9.2). For example, a phase diagram for water pinpoints the exact temperatures and pressures at which water changes from liquid to solid ice, or from liquid to steam. Similarly, metals have their own phase diagrams that indicate the exact conditions under which they change phases to become various stable alloys (such as steel from iron, or carbon steel from steel).

But mixing phase diagrams of different materials required tedious calculations, even when phase diagrams were known, because the mathematics to combine them is non-linear. Attempting to combine different metals and carbon materials to create stronger cannons became a process of guesswork. This trial-and-error method was slow and inefficient until Fourier's transform provided a way to accurately calculate the necessary conditions, greatly improving the process.

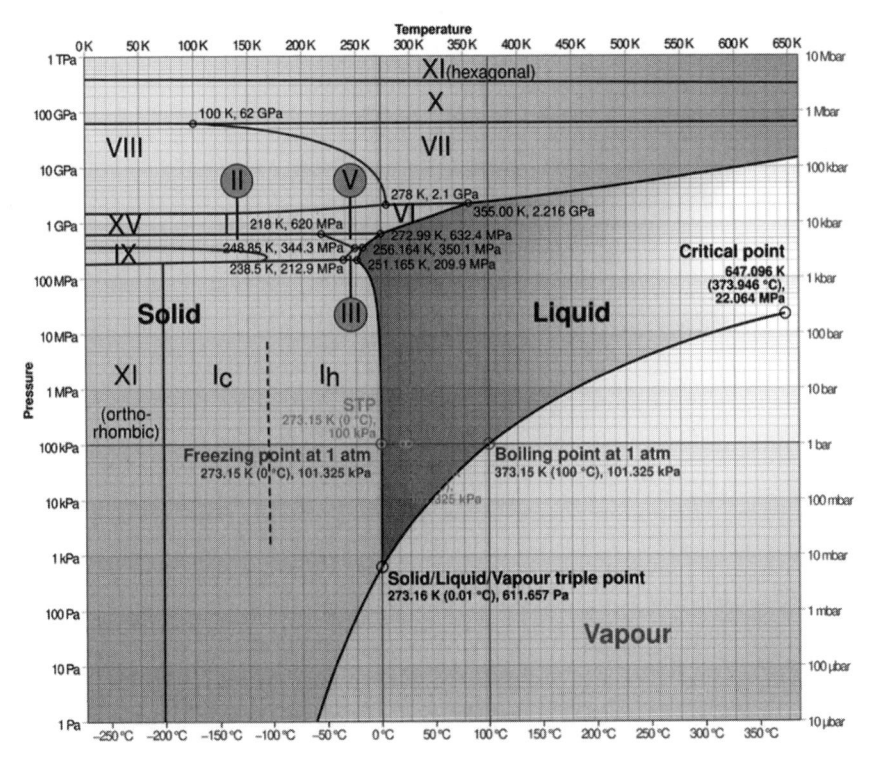

Fig. 9.2. The phase diagram of water: temperature vs pressure.

Fortunately, the same physics and mathematics that Fourier pioneered for understanding heat frequencies also apply to information signal frequencies. Fourier's equation, originally used to improve cannon casting in the nineteenth century, later became essential for enhancing data signal transmission in the twentyeth century. As engineers developed early information technology with vacuum tubes, they discovered that Fourier's equation made it much easier to manipulate and encode signals—such as by superposition, filtering, or amplification. Consequently, Fourier transform equations have become fundamental and widely used tools in physics and engineering, crucial for analyzing, synthesizing, and transmitting signals between two key domains: the time domain (td) and the frequency domain (fd).

The importance of having an equation to transform between the frequency domain and the time domain is described in the following way by Francis F. Kuo, the chief electrical engineer at the original Bell Telephone Laboratory (from which came the transistor and later, the laser). Kuo emphasizes the critical importance of an equation that transforms between the frequency domain and the time domain in his textbook, *Network Analysis and Synthesis*:

> We see that in the *time domain*, i.e., where the independent variable is t, the voltage-current relationships are given in terms of differential equations. On the other hand, in the complex *frequency domain*, the voltage-current relationships for the elements are expressed in *algebraic* equations. Algebraic equations are, in most cases, more easily solved than differential equations [. . .] Herein lies the *raison d'être* for describing signals and networks in the frequency domain as well as in the time domain.[1]

The Fourier Transform Equations

Fourier's transform equations (fig. 9.3) between the two domains of time (t_d) and frequency (f_d) are more than simply mathematical equations, written down as functions in the abstract symbolic language of calculus. The equations have revealed to scientists and mathematicians how dynamic "things" (objects, information signals, etc.) must exist in different dimen-

$$f(t) = \int_{-\infty}^{+\infty} X(F)e^{j2\pi Ft} dF \qquad f(F) = \int_{-\infty}^{+\infty} x(t)e^{-j2\pi Ft} dt$$

Fourier integral transform of a continuous frequency function into the *time domain* (t_d).

Fourier integral transform of a continuous time function into the *frequency domain* (f_d).

Fig. 9.3. The Fourier transform and inverse transform.

sions simultaneously, that the same "thing" that can be seen to exist in spacetime also exists as a doppelgänger (a "double") in a timeless, spaceless frequency dimension. Through the use of these equations, a signal (or conceivably an object) in our familiar space and time, can be converted into an identical, equivalent signal in another dimension, the pure frequency dimensions. The discovery of these relationships has had amazing implications both in philosophy and in technology. They show us how things are not what they seem, that so-called solid objects can also be seen to exist as pure frequencies in dimensions that are beyond (outside) of space and time.

These mathematical equations are widely used in modern technologies. As an example, consider how live music is recorded and then later replayed. The $f(F)$ transform is used to calculate all of the frequencies that make up the music at each sampled moment of time dt. The typical music sampling rate in modern devices is 44,000 times per second (44 kHz). The frequency data is stored or broadcast by various means. At some time in the future, the $f(t)$ transform uses the stored frequency data (dF) to play the music in the time domain.

The software to execute the Fourier transform equations is hard coded into virtually all contemporary microchips. The machine-coded software transforms any function sampled within the time domain (music, images, raw data) into a frequency spectrum that exists outside of time.

The frequency spectrum can then be fed into the $f(t)$ transform function in order to produce sound, color, and pure data when they reemerge projected into our spacetime domain. These transformations between time and frequency are the keys to our modern technological devices. The two dimensions, time and frequency, mirror one other, and it is reasonable to suspect that they may also be reflected as frequencies within dimensions that lie beyond space and time, modulated in unknown ways within the hidden, nontemporal dimensions predicted by string theory. This is explored more fully in the two chapters of part 4, "Sub-Quantum Consciousness," where they are seen reflected in Pribram's focus upon what he termed "the holonomic frequency domain" (where he suspected memory is stored), and also Bohm's concept of the implicate order as the primary source of consciousness beyond spacetime.

Common Applications of the Fourier Transform

Understanding how the Fourier transform operates in everyday applications can help clarify the principle of superposition of signals. For instance, every time an iPhone is powered on and connects to the internet, the Fourier transform plays a crucial role. Incoming frequency domain information signals are received by the iPhone's antenna (fig. 9.4) via electromagnetic waves. These signals are then quickly converted into time-based signals, such as audio and video, through the Fourier transform circuitry embedded within the iPhone. This circuitry processes signals at an impressive rate of approximately 44,000 times per second.

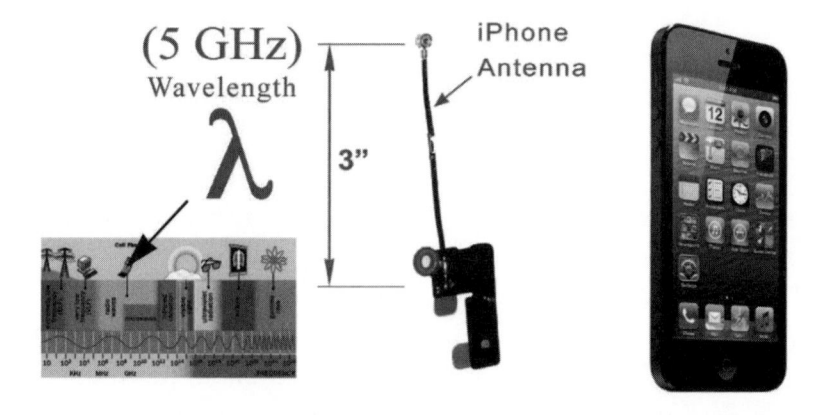

Fig. 9.4. Even iPhones have an antenna.

Another example (fig. 9.5) of the typical use of the mathematics of Fourier transforms can be found in modern digital recording and playback systems. The image shows musical audio signal vibrations coming in from the left of the figure, being picked up by a microphone in which they are converted to equivalent electrical frequencies. Next, the "sampling" of the electrical frequencies takes place. The term "bit depth" is used to indicate the number of unique frequencies that are sampled. The most common bit depths used include 16-bit for CD quality recording and 24-bit for high resolution audio. In the figure, sixteen different frequencies are shown being sampled and separated by the Fourier transform operation of the Frequency Splitter out from thousands of incoming audio frequencies. Recall that in modern recording systems, the incoming music is sampled at approximately 44,000 times per second. Each sample becomes a snapshot of the frequencies at a precise moment, and each snapshot is stored for future processing.

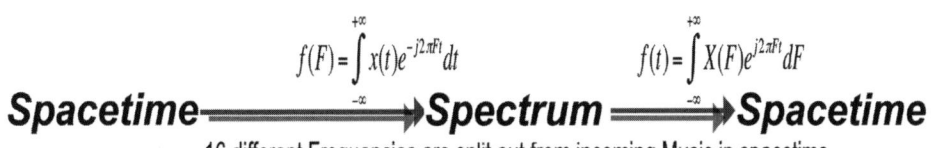

$$f(F) = \int_{-\infty}^{+\infty} x(t)e^{-j2\pi Ft} dt \qquad f(t) = \int_{-\infty}^{+\infty} X(F)e^{j2\pi Ft} dF$$

Spacetime ⟶ Spectrum ⟶ Spacetime

16 different Frequencies are split out from incoming Music in spacetime. The Frequencies are stored or transmitted. Later, a Frequency Adder transforms Frequencies into Music

Music — Frequency Splitter — Frequency Adder — Music

$$f(F) = \int_{-\infty}^{+\infty} x(t)e^{-j2\pi Ft} dt \qquad f(t) = \int_{-\infty}^{+\infty} X(F)e^{j2\pi Ft} dF$$

Fourier Transform
(Calculates each of 16 Frequencies from a signal in time)

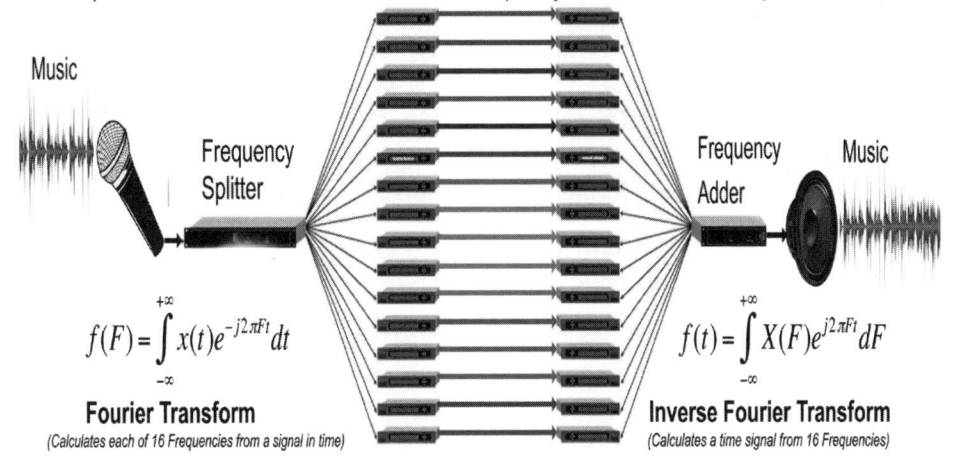

Inverse Fourier Transform
(Calculates a time signal from 16 Frequencies)

Fig. 9.5. Fourier transform of music to frequency spectrum and back to music.

Once separated, the signal strength (amplitude) of each of these sixteen separate frequencies is then recorded (on audio tape, digital memory, etc.), or directly broadcast on the internet or by radio to a location where a Fourier transform occurs within the Frequency Adder to re-combine the signals in the time domain. The superpositioned frequencies are then sent to an audio speaker where they emerge once again as the acoustical frequencies in air that can be heard as music or speech. During the process, two Fourier transforms have occurred: from time signals to frequency signals, and then back from stored frequency signals (data) to time signals (music). Not only are such equations useful for physicists and mathematicians to understand the mysterious universe, but the equations can also be used to create new ways of working with energy and information. There is also a great advantage in working with information signals in the frequency domain as the mathematics are much simpler and straightforward. By contrast, mathematics in spacetime is quite complex because nothing is linear in the geometries of spacetime, and advanced calculus and differential equation methods must be used for even simple calculations. Thus, almost all electrical engineering circuit design is carried out within the frequency domain, and only subsequently implemented with time-domain components.

Cybernetics and Fourier Analysis

Norbert Wiener (1948) coined the term *cybernetics* from the Greek κυβερνήτης— "steersman," "governor," "pilot," or "rudder"—during his work at the Bell Telephone Laboratory, making use of Fourier's transform to model and analyze brain waves in the frequency domain, where he discovered clear evidence of self-organization of electroencephalograms or brain waves. Using Fourier analysis, an approach that later became of great interest to Bohm, Wiener was able to detect uniquely narrow frequency ranges, centered within different spatial locations on the cortex, that repeatedly exhibited auto-correlation.

Wiener identified regions on the cortex of monkeys where specific ranges of frequencies were found to coalesce toward intermediate frequencies, seeming both to attract and to strengthen one another, exhibiting *resonance* or "self-tuning" to amplify and consolidate signals into narrowly specific ranges in the frequency domain f_d. Wiener's research led him to conjecture that the *infrared band* of electromagnetic flux may be the loci of "self-organizing systems":

> We thus see that a nonlinear interaction causing the attraction of frequency can generate a self-organizing system, as it does in the case of the brain waves we have discussed. This possibility of self-organization is by no means limited to the very low frequency of these two phenomena. Consider self-organizing systems at the frequency level, say, of infrared light.[2]

Three years after Wiener's 1948 publication of *Cybernetics*, David Bohm stressed the importance of Fourier's equations on the first page of his well-received 672-page textbook, *Quantum Theory* (1951), where he encouraged a familiarity with Fourier analysis for an ontological understanding of quantum phenomena: "It seems impossible to develop quantum concepts extensively without Fourier analysis. It is, therefore, presupposed that the reader is moderately familiar with Fourier analysis."[3]

For purposes of this discussion, the basic understanding of Fourier analysis coincides with Bohm's contention that *frequency vibrations* manifest within two distinct dimensions or domains: a spacetime domain and a frequency domain. Until recently, physicists have focused exclusively within spacetime to conduct their research, assuming that only space and time have any ontological "reality." Few even considered imagining that the frequency domain might lie within an ontologically real dimension, and few thought that it might be anything more than an artificial construct of a mathematical technique.

Whether there might somehow exist any "real" dimensions *outside of or beyond spacetime* has generally been dismissed or ridiculed by physicists until recently. Yet the experienced reality of transcendent dimensions of consciousness beyond space and time has been supported for centuries by the vast body of firsthand reports of religious, mystical, and near-death experiences. In an approach to such experiences, William James writes:

> The further limits of our being plunge, it seems to me, into an altogether other dimension of existence from the sensible and merely

"understandable" world. Name it the mystical region, or the supernatural region, whichever you choose. So far as our ideal impulses originate in this region (and most of them do originate in it, for we find them possessing us in a way for which we cannot articulately account), we belong to it in a more intimate sense than that in which we belong to the visible world, for we belong in the most intimate sense wherever our ideals belong.[4]

Part 4

•

SUB-QUANTUM CONSCIOUSNESS

Let me propose that consciousness is basically in the implicate order.

DAVID BOHM

10
The Pribram-Bohm Holoflux Theory

> *Currently, the terms "Dark Energy" and "Dark Matter" have surfaced as having to be measured and conceived in terms other than space and time. By analogy with potential and kinetic energy, I conceive of both these "hidden" quantum and cosmological constructs as referring to a "potential reality" which lies behind the spacetime "experienced reality" within which we ordinarily navigate.*
>
> KARL PRIBRAM[1]

There is universal agreement among modern scientists that everything that can be observed and measured in the universe of spacetime, from people to planets, is made of matter, defined as anything that has mass and occupies space. However, there is apparently much more to the universe than matter that can be seen with current scientific measuring devices. Developments in astrophysics, and particularly a stream of observations made in the 1980–90s, support the presence of what is called "dark matter" and "dark energy," that mysteriously affect and shape the cosmos. It is now thought that dark matter constitutes 85 percent of the total mass of the universe, but its origin and actual "location" remains, along with the physics of consciousness, one of the greatest contemporary mysteries in the material sciences.

While this chapter is more technical than previous chapters, the discussion here offers the reader a map of consciousness drawn from recent developments in quantum physics that may also shed light on dark energy and dark matter. Such a map can then become a framework of understanding to support an individual's efforts to explore and navigate the normally hidden dimensions of consciousness. We will focus here on the holoflux theory of consciousness as developed through the collaboration of two of the most prominent scientists of the twentieth century, the brain surgeon and neurophysicist Karl Pribram (1919–2015) and the quantum physicist David Bohm (1917–1992). For over twenty years, Pribram and Bohm worked closely in their pioneering efforts to understand and articulate a physics of consciousness that could be fully supported by Pribram's brain research and Bohm's quantum physics. Their work resulted in the articulation of the holoflux theory of consciousness.

Pribram, who has been called "the Magellan of the brain," published over 700 papers and twenty books on pattern perception, cognition, and the structure of the brain during his seventy-year career of experimental research, and he was one of the first scientists to articulate the idea that the Fourier transform might play a role in brain/mind neurophysics. Initially, Pribram spent several decades in laboratory research in his attempt to gather experimental data that might solve two problems:

1. To identify the location and mechanism of memory storage (the *engram*)
2. To discover the cognitive mechanism behind visual perception.

Pribram eventually discovered that visual signals from the rods and cones of the eyes were transformed as they entered the brain into Fourier patterns distributed over wide regions of the brain. He was able to map these fields within the fine-fibered dendritic networks of the cerebral cortex that cover the brain's outer surface.

In the mid-1960s, Pribram became fascinated by a demonstration of the first optical holograms that he saw in a trade show in Chicago. He soon realized that holograms could store, retrieve, and process vast quantities of information in holographic patterns generated by interacting frequency waves, and that this might be the way memories were stored by the brain, through the interaction of photon frequencies generated by light falling on the retinas of the eyes. In 1971, Pribram published *Languages of the Brain*, in which he detailed his new theory. He called this the "holonomic brain/mind theory," and based it upon evidence of the Fourier transform in the mind/brain process. The theory he put forth proposed that the cognitive sensory processes of memory, sight, hearing, and consciousness in general, may all operate holographically, in a transformational process of information-coded energies flowing back and forth between spacetime and the frequency domain via a Fourier transform mechanism.

Pribram's 1971 theory was radical and controversial, challenging two prominent paradigms of modern neurophysical research: (*a*) the belief that consciousness is an epiphenomenon produced by electrical sparks among synaptic clefts throughout the wiring of neurons the brain, and (*b*) the belief that somewhere in the physical brain, *engrams* of memory are stored, and will be eventually found. Pribram relates a story of a conversation he had at the time, while climbing with colleagues on a hike in Colorado just prior to attending a neuroscience conference in Boulder:

> We had climbed high into the Rocky Mountains. Coming to rest on a desolate crag, a long meditative silence was suddenly broken by a query from Campbell: "Karl, do you really believe it's a Fourier?" I hesitated, and then replied, "No, Fergus, that would be too easy, don't you agree?" Campbell sat silently awhile, then said, "You are right, it's probably not that easy. So what are you going to say tomorrow down there?" I replied, this time without hesitation, "That the transform is a Fourier, of course." Campbell smiled and chortled, "Good for you! So am I."[2]

Pribram's hypothesis was strengthened through a growing appreciation of holography as frequency-superpositioned electromagnetic wave interference. Pribram called his approach "the holonomic brain theory," and postulated the importance of the *frequency domain* in future research.

In Pribram's theory, a pure frequency domain links with the neuronal tissue of the brain through modulating fields of flux within the fine-fibered dendritic webs of the cerebral cortex regions. His paradigm was reinforced at a San Francisco conference during a lecture given by the physicist Geoffrey F. Chew, the head of the UC–Berkeley physics department and a former student of Enrico Fermi. Chew presented a conceptual diagram of the Fourier transform process (fig. 10.1), which perfectly encapsulated what Pribram had by then become familiar with, the Fourier transform.

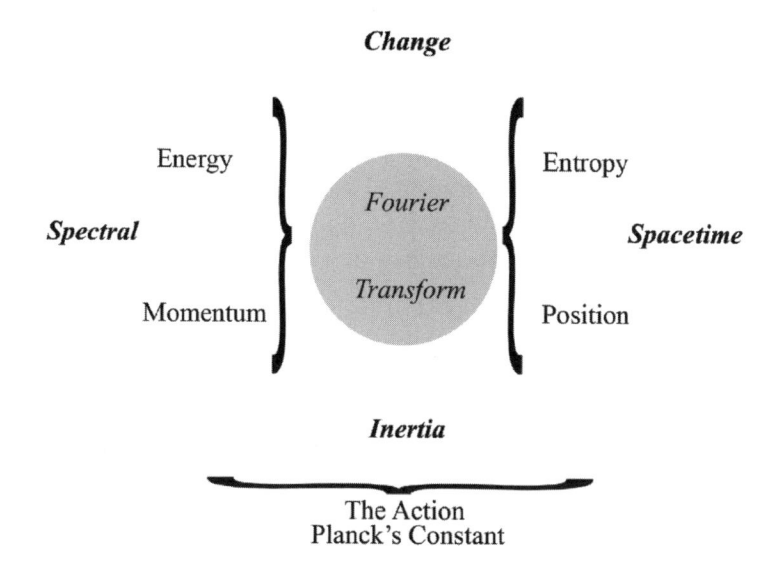

```
            Change
              
    Energy         Entropy
          Fourier
Spectral                  Spacetime
          Transform
    Momentum       Position
              
            Inertia
         ⎴⎴⎴⎴⎴⎴⎴⎴⎴
          The Action
         Planck's Constant
```

Fig. 10.1 The Dirac Fourier transform diagram.

As shown in the figure, the spectral (frequency) domain, located at the left of the diagram, is directly linked to the spacetime domain, depicted at the right, bridged by the Fourier transform, operating at the subatomic levels predicted by Planck's constant.

Pribram asked Chew where he had obtained the diagram and was told that he had been given the diagram by his colleague at Berkeley, the physicist Henry Stapp, who himself said he had been given it directly from the British theoretical physicist Paul Dirac (1902–1984), one of the original founders of quantum mechanics. Whatever the origin of the figure, Pribram chose to include the diagram in several future papers. In "Consciousness Reassessed," Pribram's caption to the figure reads, "The Fourier transform as the mediator between spectral and spacetime descriptions."[3]

In the diagram, the spectral domain is shown at the left and spacetime to the right, with the Fourier transform between them. The diagram became foundational to Pribram's understanding. It presents a two-way Fourier transform, operational at the boundary between the two domains, located at an event horizon termed in the diagram, "The Action: Planck's Constant." It is this process of turbulent transformation at the event horizon that David Bohm and Basil Hiley (1993) termed holomovement or holoflux.[4]

Pribram's breakthrough metaphor, reinforced and corroborated by the quantum physics of David Bohm, is a holographic theory of memory and consciousness: "Using the mathematical holographic process as a metaphor my (1991) book *Brain and Perception* provides detailed review of experimental results that support the conjecture that holography is a useful metaphor in coming to understand the brain/mind relation."[5]

In a series of 1989 lectures, published in his book *Brain and Perception: Holonomy and Structure in Figural Processing*, Pribram describes his holonomic theory, explaining in detail how neurodynamics aligns with quantum theory in the holographic operation of memory and mental processing.[6]

Sound, Frequency, and the Implicate Order

In contrast with the lack of success others were having in their efforts to establish the mechanism and location of memory engrams, Pribram put forth a theory that the brain mechanics of vision operates in a holonomic process, that is, through manipulation of electromagnetic holographic fields within the gray matter of the neocortex. These holographic fields are projected from a nonphysical frequency domain. The process operates through the mechanism of frequency superposition and electromagnetic wave interference.[7] Pribram called this "the holonomic brain theory," and postulated the importance of the *frequency domain*:

> Essentially, the theory reads that the brain at one stage of processing performs its analyses in the *frequency domain* [. . .] a solid body of evidence has accumulated that the auditory, somatosensory, motor, and visual systems of the brain do in fact process, at one or several stages, input from the senses in the *frequency domain*.[8]

This frequency domain has long been used in pure mathematics, but Pribram's findings assert that the frequency domain is real, and that it transcends time and space. Pribram's work shows how the mathematics of the Fourier transform may operate within the brain, modulating electromagnetic fields of flux to create, store, replay, and process holographic information that Pribram conjectures is localized within the fine-fibered dendritic webs of cerebral cortex regions.[9] According to Pribram's lab experiments, visual, auditory, and cognitive information exists within a nonspatial, nontemporal pure-frequency spectrum. This spectral information emerges into spacetime as three-dimensional electromagnetic holograms that can then be manipulated by the brain.

> It is reasonable to ask: What advantage does the organism gain by processing in the spectral transform domain? The answer is efficiency: The fact that correlations are *so easily achieved* by first convolving signals in the spectral domain and then, inverse transforming them into the spacetime domain. Thus, Fast Fourier Transform (FFT) procedures have become the basis of computerized tomography, the CT scans used by hospitals.[10]

Pribram did not base his understanding of consciousness solely on his laboratory experiments, but investigated consciousness in a direct, participatory way through his practice of meditation and mantra. In fact, it was in 1972 that he first met David Bohm at a presentation by the contemplative philosopher Jiddu Krishnamurti in Ojai, California. An even stronger indication of firsthand participatory knowledge may be found in a remarkable paragraph, written at age ninety-four, in which Pribram describes his own direct experience of "holoflux." Here, his use of the phrases "we experience" and "we can experience" strongly implies direct knowledge of such states:

> Boundaries determine what is "without" and what is "within." This raises the issue as to whether the skin can really be considered a boundary between our self and the outside world. Meditative exercises, drugs and dreams go further in dissolving these boundaries. First, we experience patterns rather than shapes and, if taken to the extreme, we can experience the loss of patterns as well: only a holoflux, "the white light" remains.[11]

There is also evidence of Pribram's association with the Transcendental Meditation movement; in 1977 a Stanford newspaper published an announcement of an upcoming presentation to be given by Karl Pribram and his Stanford colleague, William Tiller, at an event sponsored by local Transcendental Meditation organizations.[12] One of the primary teachings of the Transcendental Meditation organization, under the guidance of Maharishi Mahesh Yogi, is the practice of mantra, and it is clear that Pribram studied and practiced mantra himself.[13] Pribram specifically mentions the use of mantra in a 1991 publication, at the close of his book, in which he describes experiences of an enfolded, spectral dimension of holoflux, in a somewhat cryptic final sentence:

> Perceptual experiences may on occasion, however, reflect the spectral energy/momentum potential more than they reflect spacetime configurations: *stimuli provided by a mantra*, for example. When the spectral dimension dominates the production of a perception, space and time become enfolded in the experienced episode. Time evolution ceases and spatial boundaries disappear. The boundary between mind and matter, as all other boundaries, becomes dissolved.[14]

Pribram explained the experience of mantra stimulation in words that are virtually identical to David Bohm's description of quantum dynamics ("space and time become enfolded").

In a similar way, David Bohm's theory of consciousness was also based, in part, upon his own participatory introspection. In 1961, while a professor of physics at Bristol University, Bohm attended a conference in London to hear a series of Krishnamurti lectures on perception, a topic that, as a quantum physicist, greatly interested Bohm. During the prior year, Bohm had become interested in theories of consciousness and perception expressed in the writings of philosophers and mystics, and had begun reading books on yoga, Buddhism, and the work of the mathematician P. D. Ouspensky, a follower of the mystic G. I. Gurdjieff.[15] Public knowledge of his new interest resulted in further loss of credibility among Bohm's critics in the United States, as recounted here by his colleague and biographer, the physicist F. David Peat:

> When his former colleagues in the United States learned of this change of interest, it caused them considerable distress. In the years that fol-

lowed, some lamented that Bohm had gone "off the rails," that a great mind had been sidetracked, and the work of an exceptional physicist was being lost to science.[16]

Bohm, however, was completely serious in his search for an understanding of quantum theory, even though the quest had led him into these areas of philosophy, psychology, and mysticism.

The theory of consciousness developed by Pribram and Bohm rests upon the fundamental assumption in quantum physics that the universe in its entirety, and time and space in particular, is granular. The implication is that the foundation, the substrate of the universe, is digital in nature.

The Granularity of Space and Time: Our Digital Cosmos

When I was a young student, I was intrigued by a lecture I heard by Richard Feynman, a physics professor from the California Institute of Technology. He asked us to consider the following. If we were to take a hammer and try to strike an anvil with it, we could mathematically track the distance of the hammer from the anvil, at each moment of time. At some moment the hammer would reach half of the remaining distance to the anvil. You could plot the time of the next moment after that, when the hammer would again reach half of the distance remaining. If space were continuous, you could continue measuring the time at which the hammer managed to move closer to the anvil by halving the distance. But, if space were continuous, you would never actually be able to strike the anvil because you could continue to move half the previous distance closer to the anvil. Thus, space is not continuous, there must be a final limit, so space must be granular, not continuous. This goes for time too, he said. Time is not continuous, but granular. Then, having duly confused us students, Feynman would launch into his lecture on quantum physics.

Quantum means "discrete" in one sense, that is, there is no continuity but a sort of "jump" (in fact the phrase "quantum jump" or "quantum leap" is often used). This "jump" is due to the bottom limits to space and time that were discovered to be the solution to solve a major problem that had plagued physicists for decades, that is, the mathematics to model the relationship of temperature peaks to the peak wavelength of the electromagnetic frequency in the infrared band (that we sense and term "heat").

Max Planck (known as "the father of action") somehow intuited the solution that had escaped everyone else for decades. He had been working to solve the puzzle of the emission of radiation from heated objects, often called the problem of "black-body radiation."

All normal matter at temperatures above absolute zero emits electromagnetic radiation. It had long been noticed that charcoal could be seen to be giving off different colors that depended directly upon the temperature to which they had been heated. As the temperature of charcoal increased, the visible emission began to be seen as a dull red, then as the temperature increased, it became more orange, then yellow, then green, and finally violet. Laboratory measurements had been made of the wavelength of light coming from glowing charcoal at different temperatures (fig. 10.2), but the calculations made using "classical theory" (prior to quantum theory, shown in the dashed curve) were way off, and never matched the actual 5,000 Kelvin measurements, shown in the solid line curves. The large discrepancy for a temperature of 5,000 Kelvin can be seen in the figure.

Until Planck solved the puzzle, nobody had been able to accurately calculate the "black-body radiation curves" that would match laboratory measurements.

Planck's intuitive flash of insight was this:
1. There can be no completely continuous actions in space and time.
2. There is both a top and bottom limit to the dimensions of time and space.
3. Action must thus be "granular" (corollary: the universe is digital).

Knowing that the speed of light is invariant, Planck immediately sought to calculate the bottom limit of space, searching among various mathematical configurations to use the known constants of the speed of light and the gravitational constant in order to match exactly the

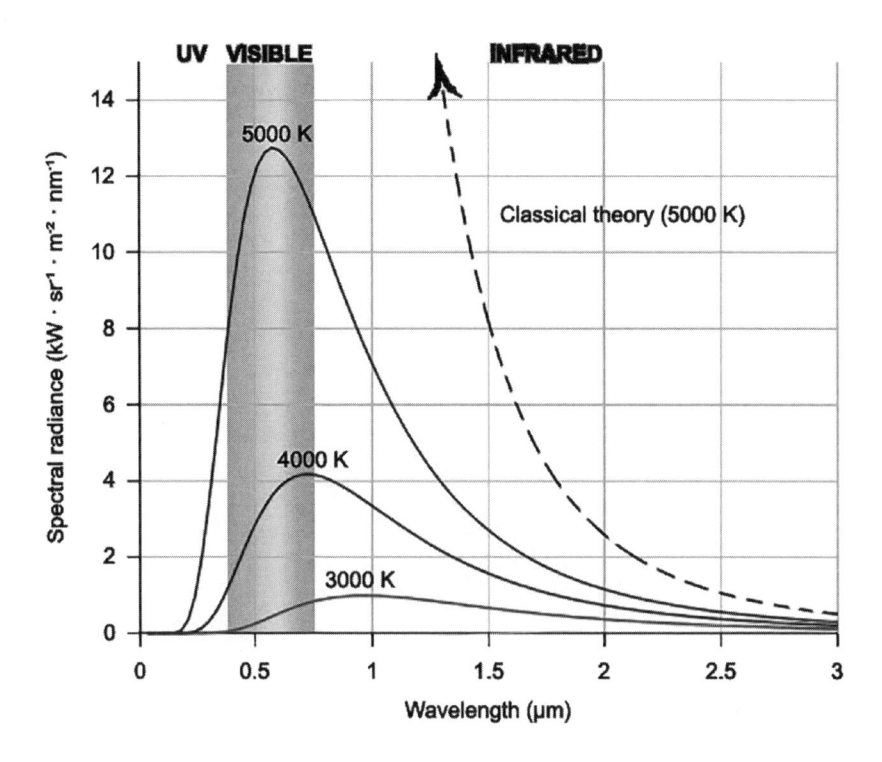

Fig. 10.2. Black-body radiation curve discrepancy.

black-body radiation curves. Eventually they all came together in his famous equation, which defines what is now called the Planck length (fig. 10.3.).

$$\ell_p = \sqrt{\frac{\hbar G}{c^3}}$$

ℓ_p: Planck length (the smallest measurable unit of length)
$\hbar$: Reduced Planck constant (quantum of action divided by 2π)
G: Gravitational constant (strength of gravity)
c: Speed of light in vacuum (maximum speed in the universe)

Fig. 10.3. The Planck length equation.

Calculation of the Planck Length and Time (universal Constants in spacetime):

$$\ell_p = \sqrt{\frac{\hbar G}{c^3}} = 1.616255 \times 10^{-35} \text{ meters}$$

$$t_p = \frac{\ell_p}{c} = \sqrt{\frac{\hbar G}{c^5}} = 5.391246 \times 10^{-44} \text{ seconds}$$

Fig. 10.4. Calculation of the Planck length and the Planck time constants.

Planck length constant	→ 1.616255×10^{-35}	Meters
Planck time constant	→ 5.391247×10^{-44}	Seconds
Planck mass constant	→ 2.176434×10^{-8}	Kilograms
Planck temperature constant	→ 1.416784×10^{32}	Kelvin

Fig. 10.5. Planck constants: granular limits to space, time, mass, and temperature.

Because length divided by velocity results in time, he was also able to calculate the fundamental unit of time (fig. 10.4) as well as several other fundamental lower limits (fig. 10.5).

These then are the lower limits to our space and time dimensions:

Note that at the smallest dimension, the Planck length, the temperature reaches the maximum possible in spacetime, the Planck temperature constant. That is because at smaller dimensions, the radii of the spinning energy "strings" is smaller, and thus the frequencies of the emitted radiation are higher. Higher frequencies emit radiation at higher energy levels.

Henceforth, using these constants in the mathematical model made the curve calculations match the observed curve exactly. A year after his discovery, Max Plank began using the term *quanta* (Latin for "how much" or "amount") to describe these values, and some years later, his colleagues began to call these the "Planck constants." In 1918 Planck was awarded the Nobel Prize for his discovery of these constants.

Comparing Various Time Units with Planck Time

	Scientific Notation	Seconds, in Standard Notation
Standard "second"	10^0	1 (note that this sets the scale of human-perceived time to approximately equal the average human heartbeat)
Early computer processor (Z80 chip)	10^{-5}	0.00001
Intel microprocessor (Intel Core i5 chip)	10^{-9}	0.000000001
Planck time	10^{-44}	0.0000000000000000000000000000000000000000001

The Planck time constant of 5.39×10^{-44} seconds is the smallest unit of time possible in our spacetime continuum. While humans are familiar with the duration of "one human second" of time, conventionally measured with our own timekeeping clock devices, it is difficult to imagine much smaller time units. One of the fastest contemporary Intel microprocessor chips runs at 4.7 GHz, which is 4.7×10^9. However, within this "one human second" there are 10^9 "processor ticks" and 10^{44} "Planck time ticks" per second, which is truly mind-boggling!

Because time does not exist below this lowest granular division of time, Planck time, there is a quantum "jump" as time flows, a jump from one tick to the next. But can we imagine what occurs between the two ticks? In between these ticks, time does not exist. This is what metaphysicians would call the void, the timeless eternal, the transcendent.

In the cosmological physics of David Bohm, it is the implicate order, the ground of being *into which* all information folds inward from our external spacetime, and from *out of which* all things are unfolded and projected into spacetime by the same implicate order.

To Bohm, all particles within our cosmos of spacetime are actually projections of a higher-dimensional order:

> We may regard each of the "particles" constituting a system as a projection of a "higher-dimensional" reality, rather than as a separate particle, existing together with all the others in a common three-dimensional space.

For example, in the experiment of Einstein, Podolsky and Rosen, each of two atoms that initially combine to form a single molecule are to be regarded as three-dimensional projections of a six-dimensional reality.[17]

Pribram and Bohm spoke about a holomovement occurring outside of the four dimensions of space and time. Pribram did not like the use of the word *movement* in this naming convention and recommended to Bohm that they use the term "holoflux," which is not a movement in time, but more of an evolution of information within the frequency domains of the implicate order. In fact one of the attributes of the implicate order, as Bohm understood it, would be a continuous *enfolding* of information from the spacetime domain, back into the hidden dimensions of the implicate order, where somehow the total information coming from spacetime for that one "tick" of time would be immediately superimposed over all of the preexisting information stored in the implicate order (Akasha, according to Bucke, Steiner, and Laszlo). At the very next quantum "tick" of time, the cosmos would be re-projected into spacetime from the information holoflux within the implicate order. The cycle can truly be seen to be a digital process, with the cosmos being created and recreated with each tick of the Planck-time clock. Bohm says:

> What follows from all this is that basically the implicate order has to be considered as a process of enfoldment and unfoldment in a higher-dimensional space. Only under certain conditions can this be simplified as a process of enfoldment and unfoldment in three dimensions.[18]

Bohm goes on to describe this projected information holoflux as it appears within the constricted spacetime dimensions:

> The electromagnetic field, which is the ground of the holographic image, obeys the laws of the quantum theory, and when these are properly applied to the field it is found that this, too, is actually a multidimensional reality which can only under certain conditions be simplified as a three-dimensional reality. Quite generally, then, the implicate order has to be extended into a multidimensional reality. In principle this reality is one unbroken whole, including the entire universe with all its "fields" and "particles."[19]

CONSCIOUSNESS

EXPERIENCE   **INFORMATION**

NON-LOCAL   LOCAL

IMPLICATE   EXPLICATE
Spectral   *Spacetime*
ORDER   *Transform*   ORDER

Holoflux Energy   **Electromagnetic Energy**
(spectrum in frequency domain)   (frequency in spacetime domain)

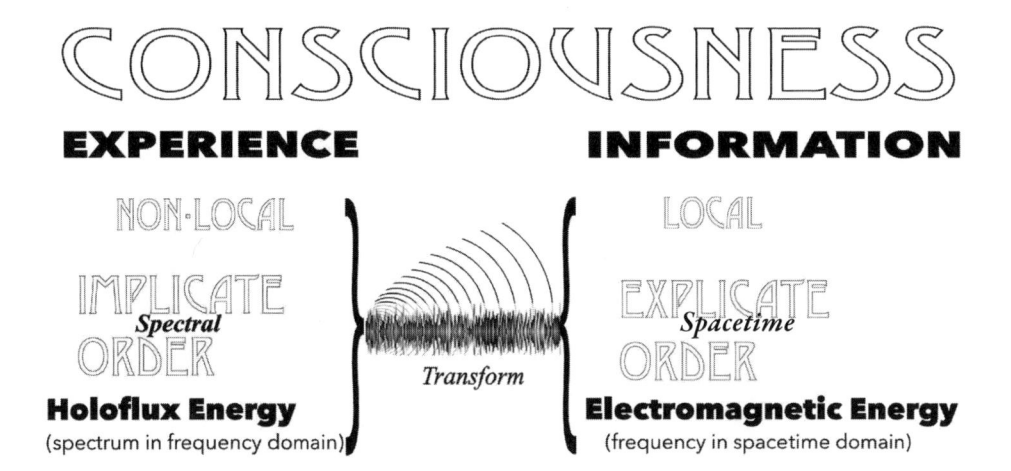

Fig. 10.6. The Pribram-Bohm holoflux model.

The Pribram-Bohm Holoflux Topology

The psychophysical model of consciousness that is developed here is presented in figure 10.6, the "Pribram-Bohm holoflux model," where the basic theory is diagrammed as consciousness transforming between nonlocal and local regions of experience and information.

The Pribram-Bohm hypothesis regards consciousness as a cybernetic energy process, a holoflux transforming between two orders of being, as Bohm said, in "an undivided flowing movement without borders."[20] To the left in the diagram, consciousness is expressed as a spectrum of holoflux energy in Bohm's implicate order. This holoflux energy resonates with electromagnetic energy of the same frequencies to the right in the diagram, in the spacetime region, or explicate order.

Viewed from left to right, the diagram reveals a spectrum of holoflux energy in the transcendental implicate order transforming and translated into "things" and "events" in local spacetime, and conversely, viewing the diagram from right to left, information generated by "things" and "events" interacting throughout spacetime is seen to be transforming (folding) back into the implicate order. The process is described as a continuous cybernetic cycle, perhaps occurring at a regular clock-rate.

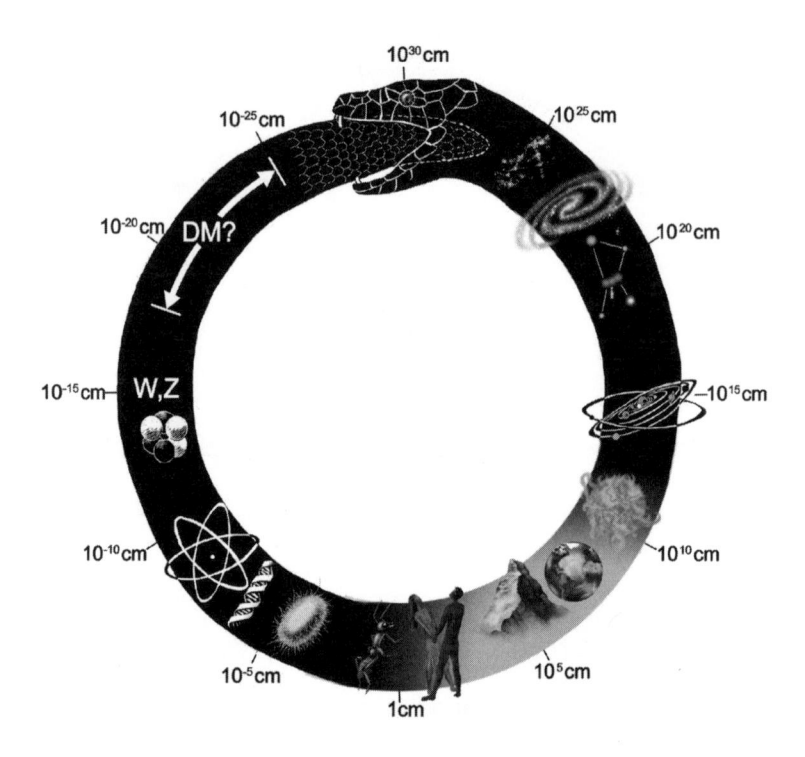

Fig. 10.7. Scales of the universe. Diagram by Bernard Carr.

The Limits of Space

From the Edge of the Universe to Planck's Constant
The Pribram-Bohm hypothesis holds that the dimensions of space are finite, and that space exhibits a limited domain in a quantifiable range. This is consistent with the physics of string theory or M-theory.

The British mathematical astronomer Bernard Carr has used the alchemical image of the ouroboros (fig. 10.7) to illustrate his grand unified theory (GUT) in comparing major scale-dependent structural levels of the physical world: "The significance of the head meeting the tail is that the entire Universe was once compressed to a point of infinite density (or, more strictly, the Planck density)."[21]

This archetypal figure implies the interconnectedness of the entire universal process in time and space, presenting a cybernetic feedback loop operational at every scale. Mystics have intuited this ouroboric process symbolized in the images of a snake swallowing its own tail (the image

The Pribram-Bohm Holoflux Theory • 227

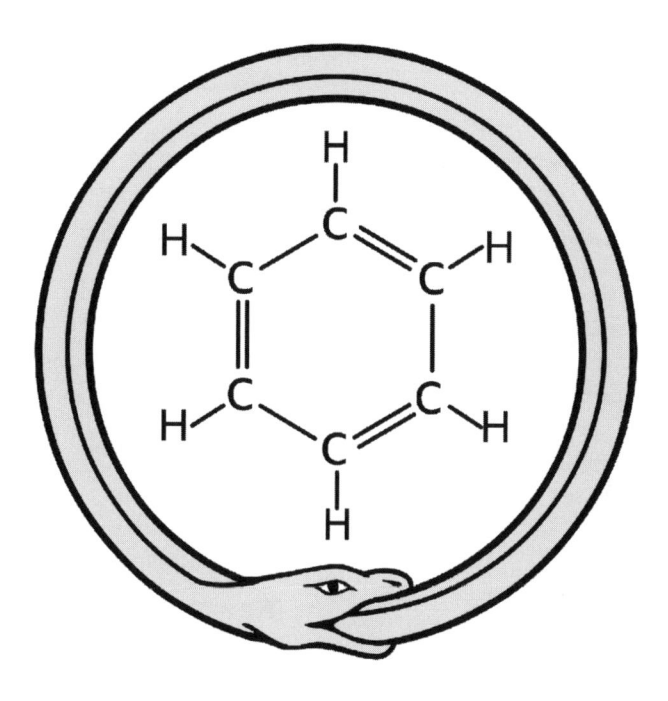

Fig. 10.8. Ouroboros.

has been found as early as the fourteenth century BCE in the tomb of Tutankhamun) and it is frequently used to symbolize cybernetic feedback in control and communication theory.[22]

One of the greatest discoveries in chemistry came as an image of the ouroboros (fig. 10.8) dreamed one night by the German organic chemist August Kekulé. In a flash of insight, that woke him from his dream, he realized that this was the structure of benzene that he had been searching for during three years of research.

> I was sitting, writing at my text-book; but the work did not progress; my thoughts were elsewhere. I turned my chair to the fire and dozed. Again the atoms were gamboling before my eyes. This time the smaller groups kept modestly in the background. My mental eye, rendered more acute by the repeated visions of the kind, could now distinguish larger structures of manifold conformation: long rows, sometimes more closely fitted together; all twining and twisting in snake-like motion.

228 • Sub-Quantum Consciousness

- 10^{25} m - Diameter of Universe
  (Astrophysics)

- 10^{20} m - Diameter of Milky Way
  (Astrophysics)

- 10^{11} m - Diameter of Earth's Orbit
  (Astrophysics)

- 10^{0} m - Human dimensions
  (Newtonian physics)

- 10^{-11} m - Hydrogen atom radius
  (Quantum physics)

- 10^{-17} m - Electron dimensions
  (Quantum physics)

↑ **Explicate order**

- - - 10^{-35} m - Planck Length - - - - - - - - - -
  (Bohmian physics)

↓ **IMPLICATE ORDER**

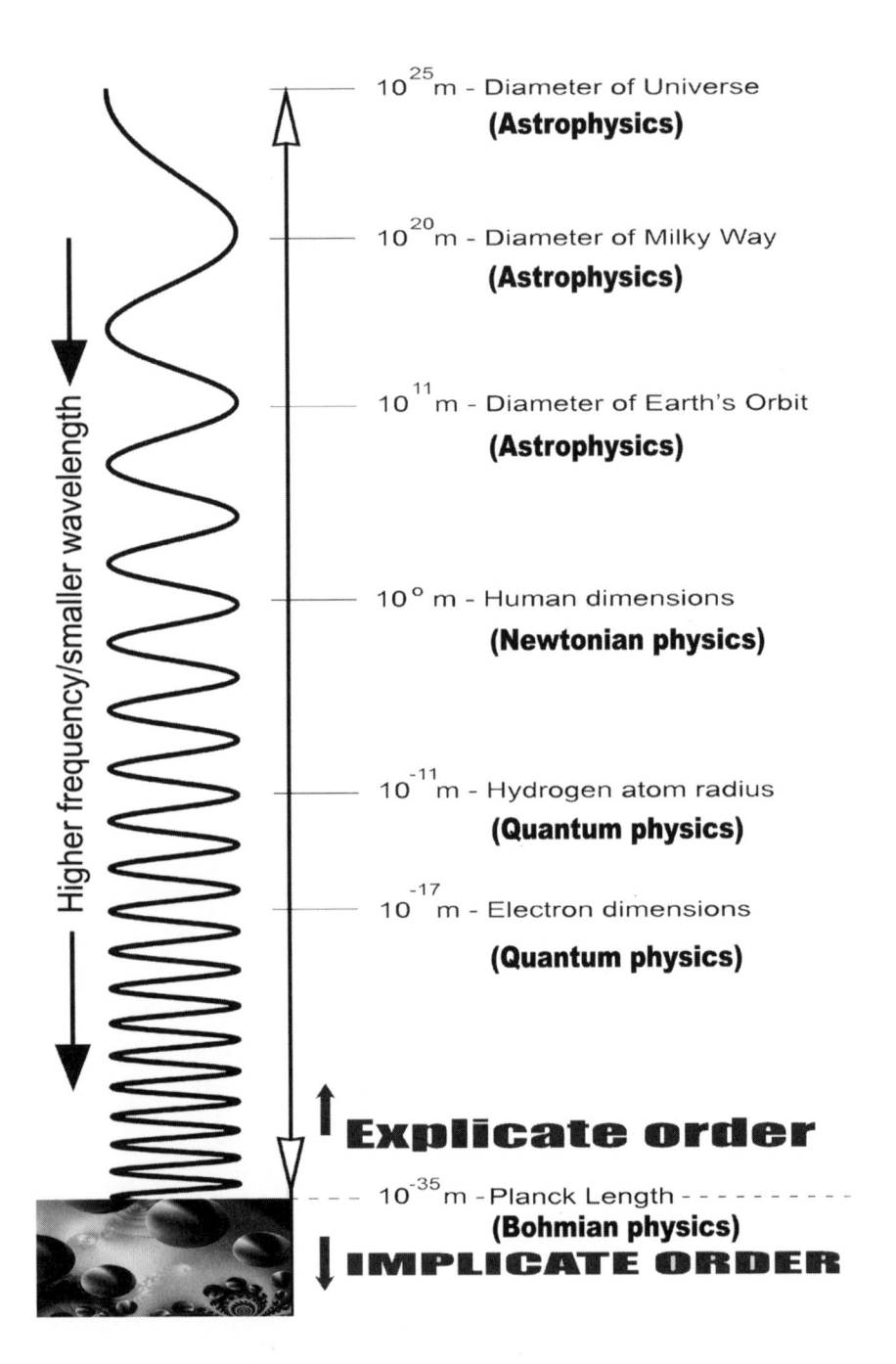

Higher frequency/smaller wavelength

Fig. 10.9. Scales of dimensional space and the explicate/implicate boundary.

But look! What was that? One of the snakes had seized hold of its own tail, and the form whirled mockingly before my eyes. As if by a flash of lightning I awoke; and this time also I spent the rest of the night in working out the consequences of the hypothesis.[23]

One can create an *axis of scales* that encompasses all of space. In figure 10.9 such a scale is drawn starting with the currently estimated diameter of the universe itself at 10^{+25} m, and descending logarithmically down to the Planck length limit at 10^{-35} m. The axis thus spans a total range of 10^{+60} (60 jumps by the power of 10). The Pribram-Bohm hypothesis holds that there, at the very bottom of the linear scale (fig. 10.9), is to be found the transition bounding the explicate order and the implicate order. Here, at the bottom bound of the spatial scale, space reaches its *end*, according to modern physics; but it also marks the *entry point into* Bohm's "implicate order," what Pribram terms the "frequency domain."

The implications of this topology are profound. Imagine moving inwardly, from any position in the universe, moving into a spherical bubble, shrinking ever smaller in scale while moving ever closer to the center at the bottom of the spatial scale, following the radial axis inward, ever shrinking downward, and then abruptly reaching the end of the line at the Planck length limit of space, the locus of a spherical shell 10^{-35} meters in diameter, below which space has no meaning. Here a boundary has been reached, an event horizon between space and the implicate order. To understand this, one must realize that the classical Cartesian assumption that space is continuous is *wrong*; there *is* indeed a bottom to space, at least according to physics, below which space no longer has meaning. Here there is a discontinuity, as David Bohm and F. David Peat explain in describing the granularity of space:

> What of the order between two points in space? The Cartesian order holds that space is continuous. Between any two points, no matter how close they lie, occur an infinity of other points. Between any two neighboring points in this infinity lies another infinity and so on. This notion of continuity is not compatible with the order of quantum theory. Thus the physicist John Wheeler has suggested that, at very short distances, continuous space begins to break up into a foam-like structure. Thus the

"order between" two points moves from the order of continuity to an order of a discontinuous foam.[24]

Pribram's Spectral Flux and the Implicate Order

In 1979, Karl Pribram, at that time a Stanford professor, attended a conference in Cordoba, Spain, where he met David Bohm, a professor of theoretical physics at London University. During the conference, Pribram soon realized that David Bohm's model of the implicate order and its projection, or extrusion into spacetime, could be seen as entirely compatible with his own holonomic mind/brain theory. Thus began twenty years of correspondence and dialog between David Bohm and Karl Pribram, and the two soon became personal friends.

Pribram saw in Bohm's theories how the frequency-domain flux might be seen to unfold into explicate-domain waves of encoded information via the Fourier transform, and he appreciated Bohm's description of how information from the explicate may fold back into the implicate in a bidirectional process. Even more intriguing was Bohm's belief that, "the basic relationship of quantum theory and consciousness is that they have the implicate order in common."[25]

Pribram was equally impressed with Bohm's explanation of nonlocality, a major mystery in quantum physics, which Bohm explains as fundamental to the process of folding and unfolding between explicate and implicate orders, allowing for full superpositioned cohesion of frequency information within the implicate order, and even providing a plausible mechanism for Sheldrake's theories of morphogenetic fields and morphic resonance.

Bohm's topology is both supported and extended by Pribram's contention, supported by the diagram handed down from Dirac, that the boundary or event horizon between the two domains, where the action occurs, is at the Planck length, precisely where, as Pribram tells us here, spectral density in-formation translates into spacetime ex-formation.

> Matter can be seen as an "ex-formation," an externalized (extruded, palpable, compacted) form of flux. By contrast, thinking and its communication (minding) are the consequence of an internalized (neg-

Change

IMPLICATE | Energy | | Entropy | **EXPLICATE**
Spectral | Fourier | | | *Spacetime*
ORDER | $f(F) = \int_{-\infty}^{+\infty} x(t)e^{-j2\pi Ft} dt$ | | | **ORDER**
| Momentum | Transform | Position |

Inertia

The Action
Planck's Constant
Event Horizon = 1.616×10^{-33} cm.

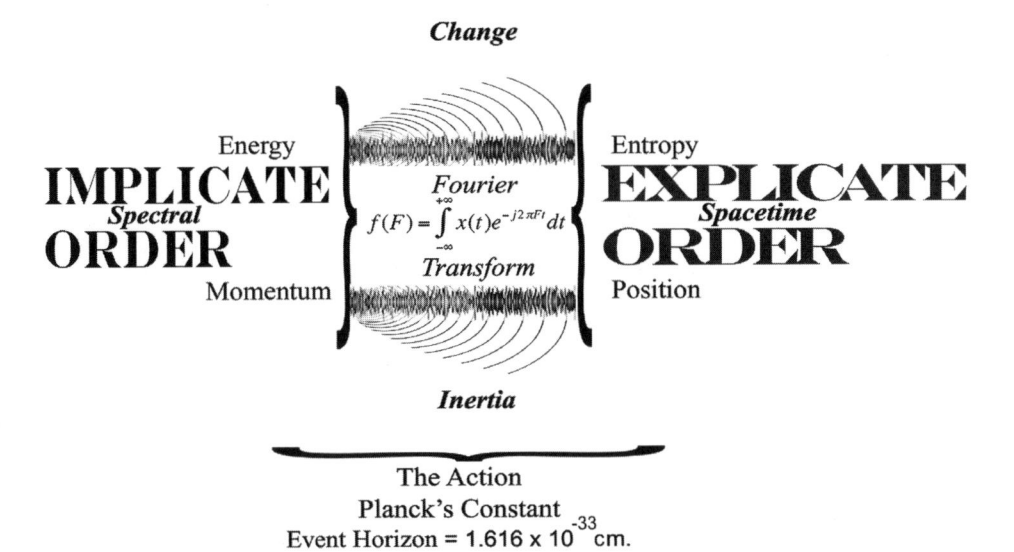

Fig. 10.10. Dirac Fourier diagram with David Bohm's topology.

entropic) forming of flux, its "in-formation." My claim is that the basis function from which both matter and mind are "formed" is flux (measured as spectral density).[26]

This flux or spectral density is for Pribram real, in the same sense that spacetime is considered to be real, but this flux is *outside of* or *beyond* spacetime. It is in this sense that Pribram made the conceptual leap from considering the Fourier transform as simply a tool of mathematical calculation, to a dawning realization that the reality of the transform implies the ontological *reality* of a domain *outside of spacetime*, a transcendent yet ontologically real domain where energy as flux is "measured as spectral density."

Dirac's original diagram can now be extended to include Bohm's two regions of the whole, the implicate order and the explicate order. Figure 10.10 shows this expanded diagram.

An anthropomorphic view of the Pribram diagram can be seen in figure 10.11. (p. 232), where an iris-like lens peering out from the implicate order is maintaining a focus upon and/or projecting a holonomic universe within the explicate order of spacetime. This mirrors Karl Pribram's conceptualization of a lens between the two domains, expressed here in *Brain and Perception*:

CONSCIOUSNESS

Frequency Domain — **Spacetime Domain**

NON-LOCAL *implicate order*

Fourier
$$f(F) = \int_{-\infty}^{+\infty} x(t) e^{-j2\pi F t} dt$$
Transform

LOCAL *explicate order*

Holosphere
(Quantum Black Hole)

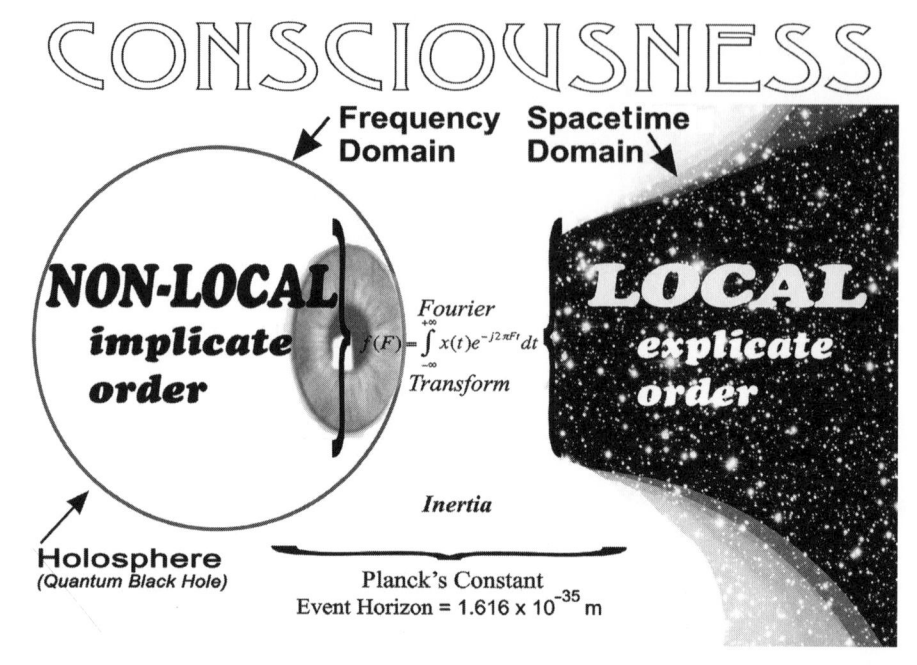

Inertia

Planck's Constant
Event Horizon = 1.616×10^{-35} m

Fig. 10.11. Topology of consciousness.

These two domains characterize the input to and output from a lens that performs a Fourier transform. On one side of the transform lies the spacetime order we ordinarily perceive. On the other side lies a distributed enfolded holographic–like order referred to as the frequency or spectral domain.[27]

Note that the image of an iris in the diagram appears at the edge of the event horizon of a quantum black hole or implicate-order holosphere. The iris symbolizes consciousness looking *out* from the implicate order *into* spacetime via a Fourier transform lensing process. This approach to a topology of consciousness as something that is looking out and seeing itself is supported here by the mathematician G. Spencer-Brown in *Laws of Form* (1972):

> Now the physicist himself, who describes all this, is, in his own account, constructed of it. He is, in short, made of a conglomeration of the very particulars he describes, no more, no less, bound together by and obeying such general laws as he himself has managed to find and record. Thus we cannot escape the fact that the world we know is

constructed in order (and thus in such a way as to be able) to see itself. This is indeed amazing. Not so much in view of what it sees, although this may appear fantastic enough, but in respect of the fact that it can see at all. But in order *to do so, evidently it must first cut itself up into at least one state which sees, and at least one other state which is seen. In this condition it will always partially elude itself.*[28]

Cosmology and the Implicate Order

In 1980, Bohm published *Wholeness and the Implicate Order*, and in a section in which he discusses the cosmology of the implicate order, he puts forth a solution to the problem of "zero-point" energy by regarding the Planck length as the shortest wavelength possible:

If one were to add up the energies of all the "wave-particle" modes of excitation in any region of space, the result would be infinite, because an infinite number of wavelengths is present. However, there is good reason to suppose that one need not keep on adding the energies corresponding to shorter and shorter wavelengths. There may be a certain shortest possible wavelength, so that the total number of modes of excitation, and therefore the energy, would be finite. When this length is estimated it turns out to be about 10^{-35} meters.[29]

Bohm brings up the school of Parmenides and Zeno, which held that all of space is actually a plenum, and he points out that as recently as the last century this same theory was presented in the widely accepted hypothesis of the *ether*. Bohm describes how there is a "holomovement" in this immense sea of "zero-point energy" to be understood as a "undivided flowing movement without borders" and he goes on to state:

It is being suggested here, then, that what we perceive through the senses as empty space is actually the plenum, which is the ground for the existence of everything, including ourselves. The things that appear to our senses are derivative forms and their true meaning can be seen only when we consider the plenum, in which they are generated and sustained, and into which they must ultimately vanish.[30]

In the Pribram-Bohm cosmology then, the interface or boundary between the spacetime explicate domain and the nonlocal, nontemporal implicate domain can be viewed topologically as a holoplenum of holospheres. Here can be found an answer to the "hard problem of consciousness" posed by Chalmers for it is from *within* each holosphere that consciousness is "peering out" into and "projecting" the spacetime explicate, and here Bohm summarizes his cosmological essay by proposing that "consciousness is to be comprehended in terms of the implicate order, along with reality as a whole" and stating unequivocally that "the implicate order is also its primary and immediate actuality."[31]

How Spacetime Projects from the Holoplenum

To summarize thus far, the Pribram-Bohm model envisions a nonlocal, transcendent mode of consciousness projecting the spatial cosmos holographically from an infinite plenum or matrix of quantum black holes, each at the limiting Planck length diameter of 10^{-35} m, located at the very bottom of space, at the very center of every three-dimensional spatial coordinate. According to quantum theory, within the implicate order, below the Planck length, space and time do not exist, as Bohm describes here:

> We come to a certain length at which the measurement of space and time becomes totally indefinable. Beyond this, the whole notion of space and time as we know it would fade out, into something that is at present unspecifiable. When this length is estimated it turns out to be about 10^{-33} cm. This is much shorter than anything thus far probed in physical experiments (which have gone down to about 10^{-17} cm or so.[32]

This leads to a new vision of the metaverse, a model in which the cosmos can be visualized as projecting outwardly from a holoplenum, from each Planck holosphere (quantum black hole) at the bottom of space, everywhere. In such a cosmology, the big bang theory would need to be revised to encompass more than simply a single point at a single time. The holoflux theory offers the prospect of multiple "big bangs" emanating from every individual Planck holosphere, perhaps a continuous recurrence cycling at Planck time (10^{-44} s) quantum ticks. Like images arising from pixels on a two-dimensional liquid crystal display (LCD) screen, the holoplenum projects a three-dimensional cosmos in all its glory. Yet during the process not only is radiant energy projecting *outwardly from* these quantum holospheres, but there is a simultaneous torrent of information, perhaps encoded in the form of gravitationally modulated energy flowing *inwardly into* the implicate order at the center of each and every point in space.

One can only imagine the transitional region between the explicate and implicate order as a frothing, turbulent, resonant event horizon that bounds spacetime and the transcendent implicate order. Information flowing inward eventually leaves spacetime entropically and flows *into* the network of quantum black holes, the holoplenum of holospheres. Information flowing into this region of infinite centers becomes immediately nonlocal, superpositioned, omni-intersecting. The holoplenum of holospheres provides a vast 3-D projection field into which spacetime is projected by implicate order holoflux.

But in addition to this spatial topology, an examination of the transition between time (in the explicate order) and timelessness (in the implicate order) is in order. How might time communicate with the timeless? What happens at the event horizon of time?

11
The Digital Universe

At the heart of quantum theory is the conviction that neither time nor space are continuous, and it has been concluded that both are granular at their lowest scale.[1] Having thus far established the granularity of space at the Planck length of 10^{-35} m, what then might be the granularity of time? One approach has been provided by the theoretical physicist John Archibald Wheeler (the originator of the term "black hole"), who, in considering a topology of information in the cosmos, noted that the highest possible clock rate must be limited by the Planck constant in general, and specifically by the *Planck time* value, below which time can have no meaning according to modern physics. The Planck time (t_P) is the time it would take a single photon traveling at the speed of light to cross a distance equal to one Planck length; the Planck length has been determined by the National Institute of Standards and Technology (n.d.) to have a value of 5.39116×10^{-44} s. Wheeler here describes the granular activity within time and space:

> Space—pure, empty, energy-free space—all the time and everywhere experiences so-called quantum fluctuations at a fantastically small scale of time, of the order of 10^{-44} seconds. During these quantum fluctuations, pairs of particles appear for an instant from the emptiness of space.[2]

Given the limit of the smallest interval of time being 10^{-44} seconds, one can calculate the maximum clock speed of the universe by determining how many Planck time intervals are possible within one human second: $(1 \text{ sec} \div 10^{-44} \text{ sec}) = 10^{+44}$

Thus, it has been proposed that if the universe is found to operate digitally, it would be cycling at its highest possible clock rate, switching between the explicate order and the implicate order cyclically at a rate of 10^{44} times per second at a bandwidth of 10^{44} Hz. This model has been developed in *The Theory of Laminated Spacetime* by Dewey.[3] The theory accords well with the Pribram-Bohm hypothesis process, and describes a process of energy flux extruding into spacetime as a series of quantum shells of information-encoded energy moving out in quantum jumps at the speed of light, with each quantum shell or brane separated from the next by a spatial gap equal to the Planck length, as described here: "I believe the universe to be composed of nothing but shells of electromagnetic particles which the theory of Laminated Spacetime describes as laminae of spacetime."[4]

Holograms versus the Holoplenum

To better understand the projection of the spacetime cosmos from the underlying holoplenum, the distinction between a hologram and a three-dimensional holoplenum projection must be drawn. A hologram is an optical image stored and recreated by a single laser (i.e., LASER: light amplitude stimulation of electromagnetic radiation), a beam of coherent electromagnetic radiation of a single frequency. Human technology greatly simplifies the creation and replay of a hologram by: (*a*) using a single beam of coherent photon energy (a *single frequency*), and (*b*) illuminating the object from only two distinct fixed points in space. An interference pattern forms from the beams of intersecting light impinging upon the three-dimensional planes of the object from different angles, causing complex shadows that are detected and recorded on the flat detector plane behind the object.

In nature, however, the situation is vastly richer in complexity: electromagnetic flux interactions in spacetime form a highly complex three-dimensional matrix of intersecting shells of every spacetime radiation frequency band conceivable, impinging from an infinity of directions. In contrast with the human generated holograms of a single frequency taken from two fixed points, each point in spacetime actually intercepts the *entire frequency flux spectrum* of the cosmos, which is being continuously captured by and feeding the event horizon entropy of the quantum black hole located at its geometric central "point."

The omnidirectional interactions manifesting among electromagnetic waves in the dimensional range of the explicate order are mirrored in the spectral frequencies within the implicate order. There is a resonance between implicate and explicate mediated by mathematical relationships such as the Fourier transforms. This process between the explicate and implicate order is skillfully articulated here by David Bohm:

> The implicate order can be thought of as a ground beyond time, a totality, out of which each moment is projected into the explicate order. For every moment that is projected out into the explicate there would be another movement in which that moment would be injected or "introjected" back into the implicate order. If you have a large number of repetitions of this process, you'll start to build up a fairly constant component to this series of projection and injection. That is, a fixed disposition would become established. The point is that, via this process, past forms would tend to be repeated or replicated in the present, and that is very similar to what Sheldrake calls a morphogenetic field and morphic resonance. Moreover, such a field would not be located anywhere. When it projects back into the totality (the implicate order), since no space and time are relevant there, all things of a similar nature might get connected together or resonate in totality. When the explicate order enfolds into the implicate order, which does not have any space, all places and all times are, we might say, merged, so that what happens in one place will interpenetrate what happens in another place.[5]

This resonance between the implicate and the explicate orders nudges into existence a cosmic holonomic metaverse of galaxies, butterflies, and zebras.

Geocentric Topology of the Holonomic Metaverse

The topology of the holonomic metaverse is congruent with cosmological intuitions of numerous classical thinkers. In the fourth century BCE, a

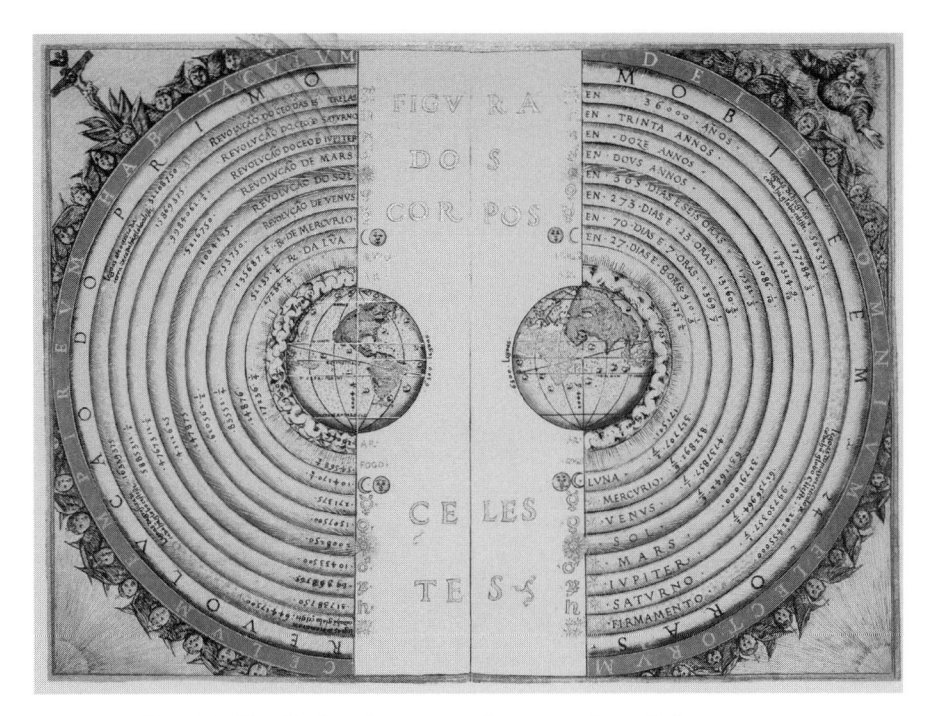

Fig. 11.1. The Ptolemaic geocentric conception of the universe. First published in *Cosmographia*, 1568.

model proposed by Plato, and further developed by his student Aristotle, consisted of a system of numerous crystalline shells rotating about a central sphere located at the center of the Earth. By the second century CE, this geocentric model had become codified by the Alexandrian astronomer Claudius Ptolemy in his cosmological model of concentric spheres (fig. 11.1).

In the third century BCE Aristotle held that the reason an apple falls to the ground is because it seeks its natural place at the center of the universe, and he set forth a geocentric model based upon the following three propositions:

1. The Earth is positioned at the center of the universe.
2. The Earth is fixed (nonmoving) in relation to the rest of the universe.
3. The Earth is special and unique compared to all other heavenly bodies.

240 • Sub-Quantum Consciousness

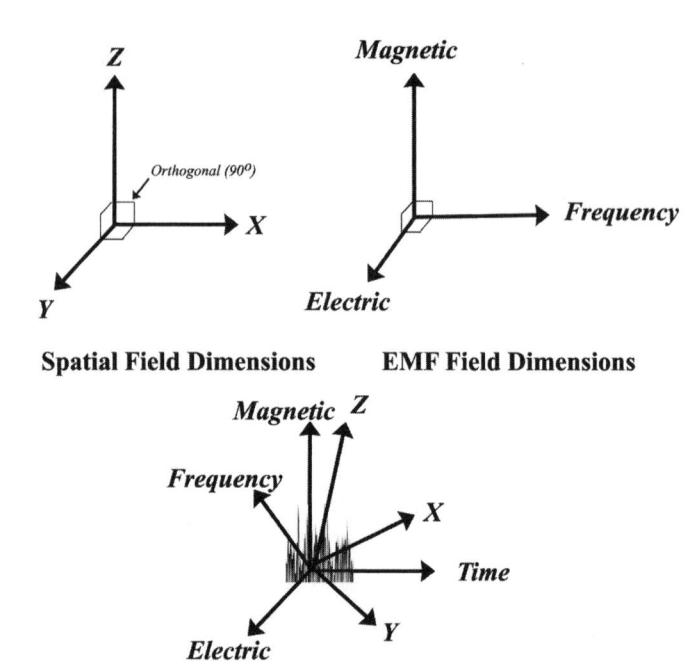

Spatial Field Dimensions   **EMF Field Dimensions**

Flux Packets in Hilbert Space

Fig. 11.2. Six-dimensional Hilbert space coordinates of a holoflux packet

Substituting "holosphere" for "Earth" in Aristotle's propositions, each Planck holosphere can be viewed as the origin of a flux packet (a coalescing region of energy in multiple frequencies and dimensions) emanating from its unique center in the universe. Each holosphere is fixed (nonmoving) in relation to all other holospheres in the holoplenum, and each holosphere is "special" by virtue of its unique Hilbert space coordinates.

A Hilbert space is a "space" having more than three dimensions. A simplified example of a Hilbert space is shown in figure 11.2; it combines the two dimensions (one is spatial, the other is electromagnetic frequency) shown in the top half of the figure. The result is a Hilbert space of six dimensions seen at the bottom center as a flux packet. A holoflux packet exists as a more complex Hilbert space of at least 11 dimensions (assuming M-theory), though possible a larger, even infinite dimensioned Hilbert space might be required to map a holoflux packet.

A first approach to a topology of the holoplenum is to begin at the event horizon of the Planck holosphere. How close to the surface of the Planck holosphere does spacetime actually begin? The overriding constraint is that it *cannot* begin any closer than one Planck length, because anything less than that has no meaning in spacetime. Accordingly, the first shell, or "isosphere," must be located exactly *one Planck length from* the central holosphere (fig. 11.3). Moving radially outward from the central holosphere, each subsequent isosphere encloses the previous isosphere. They can be visualized as nested, like hollow Russian dolls, or spherical tree rings. Within space, each isosphere is separated from the previous by a distance of 10^{-35} meters, the Planck length.

But of what do these isospheres consist? This isospheric shell cannot itself attain any appreciable depth or spatial thickness due to the Planck length constraint; they cannot be within the explicate order of spacetime while their thickness is below the Planck length. Yet if they are thinner than the Planck length, they are too thin to be in space.

Each isospheric shell is an actual event horizon, like that which separates a black hole from space and time, as first proven in 1974 with the mathematics developed by the physicist Stephen Hawking. Each isospheric shell projects ("blips" was the word Hawking used) into existence from out of the implicate order, emerging as an infinitely thin spherical shell of spacetime. However, the instant after it pops out into space, it violates the Plank length (by being thinner than the Plank length distance), and so it must immediately blip back into the implicate order. This creates, in effect, a digital signal, the clock rate of our universe, governed by the Plank time constant (that is, 5.39×10^{-44} seconds, the smallest unit of time in our spacetime universe). In the region sandwiched between each sequential shell lies the quantum vacuum (see fig. 11.3). The quantum vacuum is in actuality the region of the implicate order, the domain of every additional "hidden dimension" as suggested by string theory.

It is within each infinitely thin emerging shell itself, however, where, dynamically, the actual protrusion into space from out of implicate order exists, and this is what we perceive to be and term "space." Bohm considers the quantum vacuum that exists (outside of space), between each consecutive shell, to be the locus of the enormous energies of that which he terms the "infinite zero-point fluctuations of the vacuum field."[6]

It is interesting to note that the number of nested isospheric shells of "space" that encompass one single central holosphere (micro black hole) staggers the imagination! The number of shells can be obtained simply by dividing the size of the universe (the radius) by the Planck length to get the number of shells as shown here:

Distance to Earth from outer edge of universe [arrow] $4.40 \times 1{,}026$ m

Planck length $\rightarrow 1.6 \times 10^{-35}$ m

number of shells = $(4.40 \times 1{,}026 \text{ m}) \div 1.6 \times 10^{-35} \text{ m} = 2.7 \times 1{,}061$

A topology of the implicate order is here coming into focus. A nonlocal, timeless, transcendent domain is to be found emerging into space as nested, infinitely thin spherical shells of energy projecting from a central holosophere of the implicate order. We might imagine that upon the surface region energy flux of each of these 1,061 isospheric shells of space that emanate from every point in our body, may be stored holographic information. Each isosphere would thus provide a structural capacity for information storage limited only by the Bekenstein bound (discussed in the next section).

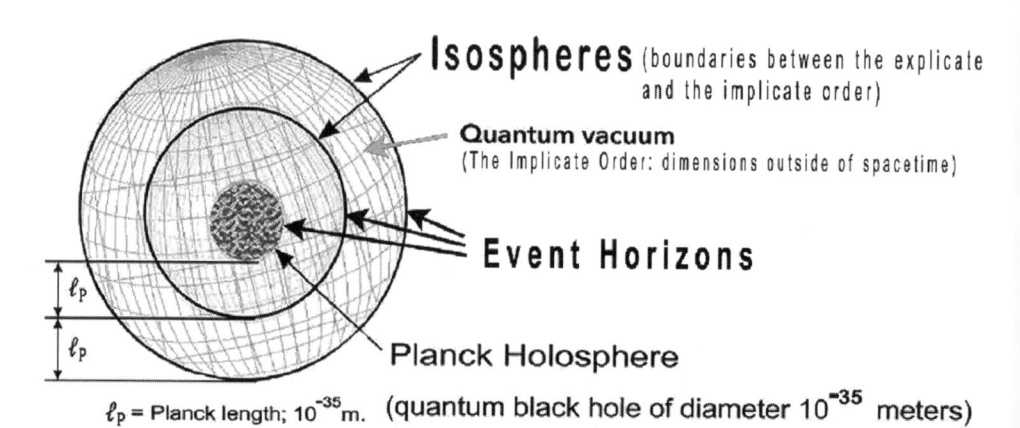

Quantum Shells of the Implicate Order

Isospheres (boundaries between the explicate and the implicate order)

Quantum vacuum (The Implicate Order: dimensions outside of spacetime)

Event Horizons

Planck Holosphere

ℓ_P = Planck length; 10^{-35} m. (quantum black hole of diameter 10^{-35} meters)

Fig. 11.3. Quantum shells of the implicate order.

While distributed geometrically like nested Russian dolls these isospheric quantum shells of emerging and disappearing space would share in the nonlocality of the implicate order from which they project, thus providing both a fundamental structure for storage of memory and the requisite information backbone or lattice that bridges the hidden dimensions with our spacetime universe.

As in all black-hole phenomena, these isospheres (as well as the central Planck holosphere) are assumed to have an angular momentum (spin) due to turbulence at the event horizon, a highly energetic region that, the holoflux theory posits, are modulated with qubits of information. These spinning, implicate order isospheres project spacetime "objects" (multidimensional holograms) from frequency-phase information within the implicate order.

Spectral frequencies within the implicate order (the frequency domain of communication engineering) can be correlated with sets of isospheres having explicate-order diameters corresponding to wavelengths of the same frequency. Thus a complex patterned frequency spectrum within the implicate order can be mirrored in the spatial dimension of the explicate order through the activation of selected ranges of shell isospheres of the implicate order.

Having established the mechanism of communication linking Bohm's implicate and explicate orders, it is now possible to turn to memory storage, that elusive problem that first led Pribram to his holonomic theory of a brain/mind interface.

A Holoplenum of Quantum Black Holes

As described in the previous section, a holonomic process underlies the Bohmian "Whole" and operates through means of a cosmic plenum of quantum black holes. These quantum black holes present as geometric shells or spheres in space, each one being a boundary between space (outside of the sphere) and the hidden dimensions of the implicate order within the shell boundary (though "within" is a spatial metaphor). Each of these quantum black holes is exactly one quantum in diameter, that is, one Planck length. Thus these quantum-diameter shells have been termed "Planck diameter holospheres."

In geometric terms, these "close-packed spheres" refers to the most tightly packed and highly efficient arrangement of spheres of equal diameter, each of which touches multiple similar adjacent spheres. The entire geometric plenum can be visualized at the very bottom of space, existing everywhere in our spacetime universe, and considered metaphorically as a three-dimensional "ocean floor" of space.

How Holopixels Project the Universe

These inconceivably tiny holospheres act as pixels, or *holopixels*, continually projecting outward into our four-dimensional universe of spacetime from the hidden dimensions of the metaverse. Much in the same way that the two-dimensional array of pixels on the back of a plasma display screen projects a streaming video image that we can see, this plenum of quantum holopixels projects the universe into our space and time dimensions. Yet, the region within the boundaries of these black holes contains the multiple hidden dimensions beyond spacetime.

This three-dimensional substrate or plenum exists everywhere and continually projects our spacetime universe that we call the cosmos, as depicted in this alchemical wood engraving (fig. 11.4).

In totality, this plenum of holopixels acts much as the two-dimensional plenum of pixels that are embedded in the substrate of a modern large-screen plasma video display. Surrounding and radiating outward from each central Planck holosphere throughout space are spherical shells of quantum potential, infinitely thin information-encoded shells of energy, each separated from an inner shell by the quantum exclusion radial distance of one Planck length. These isospheric shells of Bohmian quantum potential extend out to the diameter of the universe itself.

Thus can be visualized topologically an almost infinite series of nested isospheric shells, spread out at discrete quantum radii from their respective central Planck holospheres, each bounding Bohm's nondual implicate order.

The global panoramic intersection of these holospheric shells manifests as the projection of the holonomic universe into spacetime. The cumulative effect of this projection, as regarded by physicists observing from significantly higher scalar dimensions, is described as "matter." The phenomenon can be understood as a *process of projected creation*, an omnipresent, ongoing holographic extrusion of information *from* the implicate order *into* the explicate order, where the various structures of the cosmos

Fig. 11.4. A seer at the intersection of heaven and spacetime.
Graphic by Camille Flammarion (Paris 1888).

(galactic clusters, stars, etc.), the complex unfoldings *into* spacetime, are perceived by the human eye and mind to be three-dimensional, when they are actually holographic projections *from* the implicate order, from the center outward.

Another way of visualizing the projected illusion of spacetime reality from the holoplenum is by expanding upon the metaphor of a flat-panel plasma display (such as the one you may be viewing as you read this).

Consider the human visual threshold for detecting separate images, which lies somewhere between ten to twelve images per second; the industry standard in the motion picture industry is twenty-four frames per second. This standard ensures that the presentation of a sequence of projected images will appear to a human viewer as a smooth and continuous motion.

By contrast, if the entire universe flashes in and out of existence at a clock-cycle rate limited only by the Planck time constant of 5.3×10^{-44} seconds, equivalent to a "frame" rate of almost 10^{44} "frames per second," the cosmos would *appear* to be smooth and continuous in all respects even to an electron, and certainly to any human observer of the cosmos, even at quantum dimensions.

The approximate image resolution of a "holoplenum display" obtained by dividing one inch by the Planck length, yields a maximum resolution of 1.584×10^{34} holopixels per inch. At such hyperfine resolution, even a Higgs boson in the 10^{-17} meter dimensional range would appear to be moving smoothly through space.

Holonomic Storage: The Bekenstein Bound

In *Wholeness and the Implicate Order*, Bohm articulates and develops a "quantum potential" function that projects the explicate spacetime universe out from within an enfolded sub-quantum implicate order.[7] Bohm's quantum potential function is congruent with de Broglie's "pilot wave" theory of 1927, as both are based upon a conviction that there exist "hidden variables" in sub-quantum regions not accessible to observational exploration using current material-science technology (and far beyond the capabilities of the CERN Large Hadron Collider).

The de Broglie pilot wave theory and Bohm's quantum potential are mathematical attempts to map sub-quantum effects issuing from an implicate order in a domain of "hidden variables" far below the observational capabilities of contemporary material science. Both theories posit a cybernetic processing of information, simultaneously being cycled from the spacetime world and enfolded into the nondual frequency domain where the accumulating information is processed nonlocally within the implicate order. Driven then by the implicate order, a pilot wave of quantum potential nudges the configurations in spacetime into an altered, slightly new configuration, much as a small tugboat might influence an enormous freighter. If the cosmos operates at its maximum possible clock cycle, as discussed previously, this pilot wave might be seen to operate at the extreme clock-cycle rate of the Planck time constant, or 10^{44} Hz.

Somewhere, however, such a cosmic process would require a memory storage repository in spacetime. Regarded as a cybernetic process, the sequence of information feedback and action can be metaphorically imaged in the alchemical ouroboros, the classical symbol for consciousness, depicted as a snake in a circular configuration eating (or chasing) its own tail. This process can be viewed as the cyclic transfer of information coming in (from the tail) and the resulting action (by the head). A cybernetic feedback loop thus needs data, information, as input. Where then might data be accumulated and retrieved in spacetime at these most fundamental sub-quantum levels, in a Bohmian holonomic universe consisting topologically of the distributed plenum of Planck holospheres, each surrounded by a series of nested quantum isospheres?

One possibility is to consider the information storage potential of an isosphere encoded with granular "bits" of data. In 1970, Jacob D. Bekenstein, then a graduate student working under John Archibald Wheeler, proposed a novel idea. Bekenstein proposed that there must be an absolute maximum amount of information that can be stored in a finite region of space, and that the Planck constants in quantum theory can be used to determine this limit. Twenty years later, Bekenstein's theory was extended into what is called the *holographic principle* by Leonard Susskind, which describes how information within any volume of space can be encoded on a boundary of the region.[8] A description of this configuration is presented here by Wheeler himself, as first related to him by Bekenstein:

> One unit of entropy (information), one unit of randomness, one unit of disorder, Bekenstein explained to me, must be associated with a bit of area of this order of magnitude (a Planck length square) [. . .] Thus one unit of entropy is associated with each 1.04×10^{-69} square meters of the horizon of a black hole.[9]

This proposed upper limit to the information that can be contained upon the surface of a specific, finite volume of space has come to be known as the "Bekenstein bound."[10] Symbolically depicted in figure 11.5 is a topological depiction of the arrangement of information bits, or "qubits," stored on the bounding surface of a spherical volume or isosphere.

This same topological approach to data storage can be applied to human physiology. Using a well-known biological structure as an example, it is possible to calculate the maximum memory storage capacity of an isosphere the size of a single erythrocyte, the ubiquitous red blood cell found

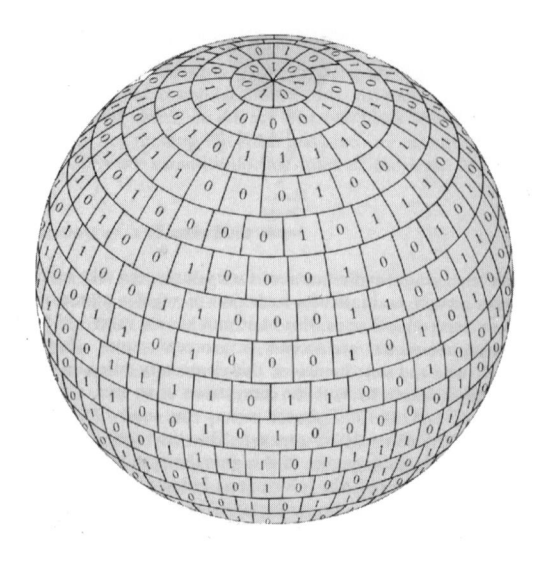

Fig. 11.5. Planck length qubits on the surface of an isosphere. Image from Wheeler, *A Journey into Gravity and Spacetime*, 220.

throughout the human body. Using Wheeler's approach to determine the number of Bekensteinian equivalent data bits (qubits) on the surface of an isosphere, and using the average diameter of a typical human erythrocyte of 8.1 microns (or 8.1×10^{-5} m), the maximum possible storage capacity on the surface of a single red blood cell can be calculated.[11] To obtain this limiting number of bits, the surface area on a spherical shell 8.1 microns in diameter must be divided by 1.04×10^{-69} square meters (which is the Bekenstein unit of entropy, or approximately the square of the Planck length of 1.616199×10^{-35}). The surface area of this erythrocyte-bisected sphere according to this calculation is or $4\pi*(8.1 \times 10^{-5})^2 = 4\pi*(6.561 \times 10^{-9}) = 8.24 \times 10^{-9}$ square meters. Dividing this by the qubit area of 1.04×10^{-69} square meters yields an estimated maximum storage capacity of 8×10^{60} qubits of storage space for potential information encoding. This is an extremely large data storage capacity, considerably larger than, by contrast, the entire projected capability of the National Security Agency's Utah Data Center, which has been designed, when completed, to have a maximum data storage capacity of twelve exabytes or 12×10^{18} bytes.

Unfolding the Implicate Order into Spacetime

The ontological understanding of quantum physics that Bohm sought emerges in this model of a sub-quantum, omnipresent holoplenum of Planck holospheres, each enclosing a contiguous transcendent region of nonspatial, nontemporal dimensions termed by Bohm "the implicate order."

What are the implications of this model for human consciousness, cognitively operational at temporal and spatial scales vastly larger than those found at these sub-quantum Planck boundaries? To answer this we must first complete the Pribram-Bohm cosmological topology of consciousness, and to do this the concept of isospheres, shells within shells, must be considered. Moving outwardly, radially, from the interior bounding event horizon at each central Planck holosphere, can be identified isospheric shells of the implicate order, extruding into space at exact Planck length (quantum) intervals.

This series of concentric shells, each one separated from the next by one Planck length, are isospherical loci of the implicate order (see fig. 11.6); they extrude into space and they intersect in space with other shells bounding other Planck holospheres in the spacetime holoplenum. It is the cumulative interference effect of the intersection of individual isospheric shells that project images at higher scales, holographically, into three-dimensional space.

Isospheres

The Quantum Black Hole/ Planck holosphere (diameter 10^{-35} meters)

Bottom (lower) Limit of Spacetime domain

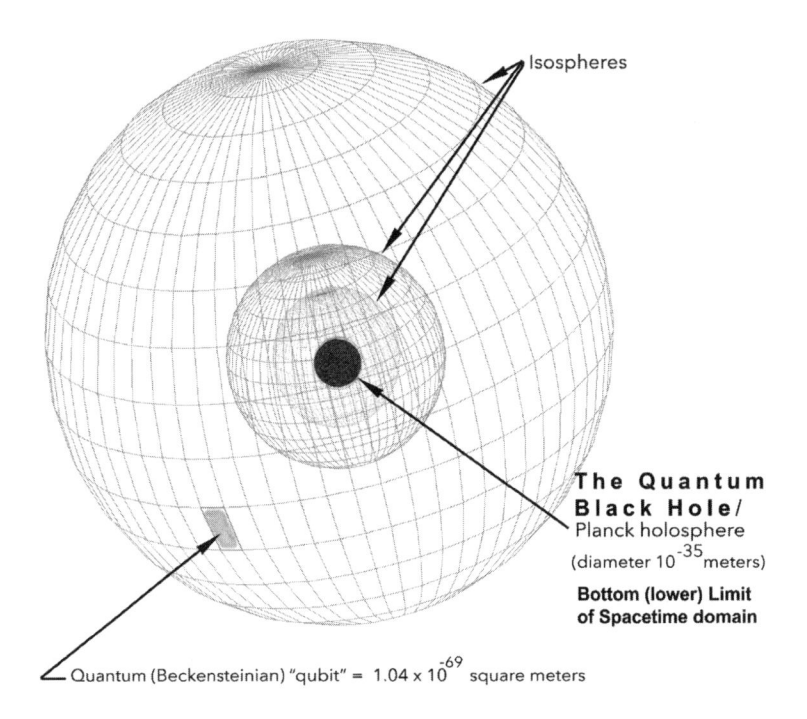

Quantum (Beckensteinian) "qubit" = 1.04×10^{-69} square meters

Fig. 11.6. Topology model of isospheres surrounding a Planck holosphere.

Each isospheric shell, as Bekenstein determined, has a potentially enormous storage capacity in "qubits" of information encoded on the event horizon bounding each shell, depending only upon the radius of the shell within the range of 10^{-35} m to 10^{27} m. Accessible simultaneously in both the implicate order and the explicate order, such encoded information provides the data to guide evolving forms as they project into the explicate via the pilot wave mathematics of Bohm's "many-dimensioned quantum potential." As part of this process, in-formation becomes ex-formation as the implicate order unfurls into the explicate order. The plasmoidal forms appearing in spacetime as electromagnetic flux energy are mirrored by and resonate within the implicate order as the dark energy of frequency-phase holoflux.

Within this cosmic geometry of the Pribram-Bohm hypothesis can be identified a framework for omnipresent two-way portals, potential bridges between the explicate and the implicate. Here, the deep consciousness of the universe flows in a cyclic, cybernetic, perhaps fractal movement in time and space, moving through an endless, bidirectional process, consciousness involving itself in a dance of transformation.

How then does this topology support human consciousness, thought, and perception in spacetime? How can electromagnetic frequency plasma in spacetime resonate with holoflux plasma in the implicate order? First of all, the energies must be within the same frequency range in order for maximum interactive resonance to occur. Where the frequencies overlap as they superimpose and interpenetrate one other, resonance occurs.

Resonance is a naturally occurring phenomenon characteristic of physical objects or plasma fields extended in spacetime. The resonance effect is seen when objects or complex signal systems exhibit remarkable frequency sensitivity to particular external frequencies, flowing in, through, and around the system, frequencies that approach the "natural resonant frequency" of the object or signal system; perfect resonance occurs where the input frequency and the natural frequency are identical.[12] This principle of resonance governs all cybernetic feedback loops, and is a key factor in the design of antennas for electromagnetic transmitting and receiving systems. The goal of antenna design is to construct an antenna that is maximally resonant within a specific narrow frequency range of incoming (external or internal) electromagnetic radiation; when the input frequency and the natural frequency of the antenna coincide or move significantly close to one another, resonance occurs. This simplest and most common antenna design is the dipole antenna, which depends directly on the size (wavelength) of the incoming electromagnetic wave. A dipole antenna is designed to be physically half the size of the incoming wavelength. This half-wavelength effect governs the design of network communication waveguides applied to fiber optics in the internet, where the antenna is the fiber channel itself acting as a waveguide. The waveguide is highly efficient for two reasons: first, as its name implies, the waveguide guides the electromagnetic wave within its channel with maximum efficiency; and secondly, it shields the signal in the channel from external electromagnetic waves. The inner diameter of the hollow waveguide is designed to be equal to exactly half the wavelength of the electromagnetic energy signal shielded by and flowing through the waveguide channel.

Waveguides in the Human Capillary System

Waveguides have been used for over a century both commercially and in research to channel and guide vibrating energy of specific limited frequency ranges; the fiber-optic networks hosting the global internet operate on this principle, channeling electromagnetic radiation at fixed laser frequencies.[13] It was discovered late in the nineteenth century that circular metallic tubes, or hollow metal ducts similar to A/C ventilation ducts (but much smaller) could be used to channel and guide sound vibrations in air, or electromagnetic energy in the air, vacuum, or a transparent solid such as is used in fiber-optic cables that connect the internet worldwide.

Without the waveguide, the vibrational energy field would be transmitted in all directions, visualized as magnetic lines or arrows emerging from a point at the center of an expanding sphere. This energy would be dispersed outwardly, the magnetic vectored arrowheads might be seen to be pushing out spherically in every direction from the central point of origin.

A waveguide, however, constrains the magnetic components of the wavefront of vibrating energy to one specific linear direction, in parallel with the center of the waveguide, and thus, conceptually, the confined wave itself loses very little power while it propagates directly along the central axis of the waveguide, like a stream of water emerging from the pinprick of a large, taut, water balloon.[14]

The most common type of fiber-optic cable used in the internet has a core diameter of 8–10 micrometers and is designed for use to send and receive signals in the near infrared band. Fiber-optic cable is thus designed for use in the infrared band of wavelengths.[15] These encoded wavelengths of infrared light make up the global brain that we call the internet, through which signals runs at the speed of light across the surface of the planet. This global fiber-optic network is powered by highly efficient carbon dioxide lasers that use carbon dioxide molecules that resonate at a wavelength of 10 μm (microns). Is it a coincidence that the human blood capillary diameter happens also to be 10 μm? Or that human blood capillaries are full of carbon dioxide? Or that diameter of a human brain neuron is also 10 μm, ideal to act as an infrared waveguide? Are we to assume these are all coincidences?

The average human blood capillary (also 10 microns in diameter) is at all times full of blood rich with carbon dioxide. Given these facts, is it not far-fetched to imagine each human blood capillary system acting very much like our modern fiber-optic laser communication systems by providing the waveguide and carbon dioxide laser capability to maintain a high-speed internet-like communication system. It appears that Jung was definitely on the right track in viewing consciousness and the psyche operating across a wide energy-frequency spectrum.

It should also be noticed how the experience of time itself can be seen to be dependent on the relative spatial scales of spacetime. Time experienced by a human may be quite different from time experienced by a paramecium, or by the sun, or a galactic-sized psychoid.

Dimensional analysis and a cursory examination of human physiology would immediately suggest two candidates for waveguide systems within the human body: (*a*) the blood capillary system, and (*b*) the microtubule system. The corresponding resonant frequency for electromagnetic waves using such waveguides corresponds to wavelengths matching the inner diameter of these structures. For blood system capillaries, this corresponds to radiation with a wavelength of 9.3 to 10.0 microns, the average inner diameter of a capillary. For microtubules, the radiation wavelength would be found in a range of 40 nanometers, the inner diameter of the microtubule waveguides. Figure 11.7 (p. 253) depicts the location of each of these potential waveguide frequency bands within a wider section of the electromagnetic spectrum.

There is no impediment to our blood acting as an electromagnetic plasma within the capillary system, and, as previously mentioned, the opening page of a textbook on plasma physics reads, "It has often been said that 99% of the matter in the universe is in the plasma state."[16]

In such a model, the entire blood system within the human body can be considered to act as an extensively polarized "super cell" of nonlocal

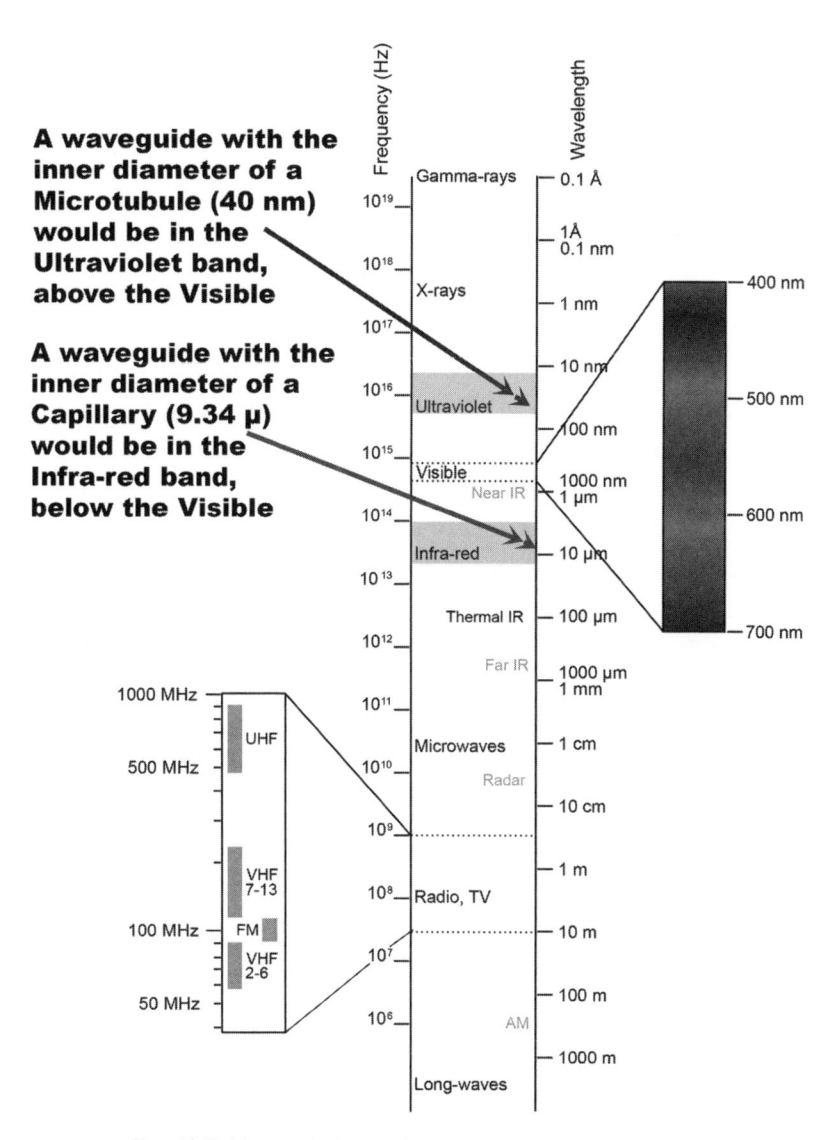

Fig. 11.7. Microtubules and capillaries as waveguides. Annotations by author; graphic by Jahoe.

electromagnetic plasma energy, which can then be differentiated from the neuronal brain body of consciousness, itself generated by sequential electrical impulse-driven patterns flowing in the nervous system. Moving charges generate magnetic fields, and ionized human blood flow is no exception: flowing blood plasma results in creation of a magnetic field, and this is in accord with the conjecture of moving charges generate magnetic fields, and ionized human blood flow is no exception: flowing blood plasma results in creation of a magnetic field, and this is in accord with the conjecture of Quantum Brain Mechanics (QBD).

The circulatory system can be seen as a magnetic plasma composed primarily of ionized red blood cells (erythrocytes) and water molecules, flowing together in complex vortices of blood plasma around every cell and through every capillary of the body. Each erythrocyte is a flexible, annular, bioconcave disk shaped like a doughnut (in geometry, a torus), having a thin webbed center where the hole in a pastry doughnut would be located. The typical outside diameter of a red blood cell is approximately 9 microns, close to the infrared wavelength of 9.6 microns generated by the human body. The adult human body contains approximately 6 grams of iron, of which 60 percent is stored throughout the 10^{12} erythrocytes, each of which contains approximately 270 million atoms of ionic iron embedded within transparent hemoglobin in a toroidal locus. Thus, each erythrocyte, replete with iron ions embedded in hemoglobin, creates in effect, an ionized iron toroid.

Recent studies have also discovered neuronal generation of electromagnetic energy in the near infrared region of the spectrum centered around 10 microns. Radiation emission was repeatedly measured emanating from live crab neurons in extremely narrow, discrete spectral bands within the frequency range corresponding to a spectral region from 10.5 to 6.5 microns.

The implications of this model are considerable: there may exist in nature a unique resonant frequency for each individual human being. It is useful here to step through a topological analysis of the possible functions of a human red blood cell, given its geometry, as a locus of consciousness, and the possible use of the erythrocyte as a locus of memory storage at human biological scales. If, as previously conjectured, the red blood cell has an ideal diameter to resonate electromagnetic radiation in its ferrite-embedded ring at the human infrared wavelength of 10 microns, then it is reasonable to ask if this configuration could accommodate a single unique isospheric frequency (wavelength) for each of the currently eight billion living humans on the planet. In other words, does this geometry allow for the possibility of each human being to also have a single unique frequency within the infrared electromagnetic radiation that resonates within the human cardiovascular waveguide system? Figure 11.8 outlines the topological feasibility of this approach.

At what wavelength (and temperature) then might the ego operate in Jung's proposed spectrum of consciousness? The answer can be determined by human body temperature, known to peak around 98.6°F, a temperature at which the emitted radiation resonates at a wavelength of 10 microns that is observed to be in the infrared region, just below the red wavelength range in the visible light spectrum. Coincidentally, the exact wavelength generated at 98.6°F perfectly matches the inner diameter of human blood capillaries, which range from 8 to 10 microns, found throughout the body. By adulthood, the average human body contains approximately 60,000 miles of capillaries, a length that is more than twice the distance around the planet.

How then might consciousness operate within the human body? From the information technology point of view, the capillaries can be seen to act as waveguides for infrared electromagnetic signals streaming throughout the human body, carrying signals in a vast network of two-way information processing.

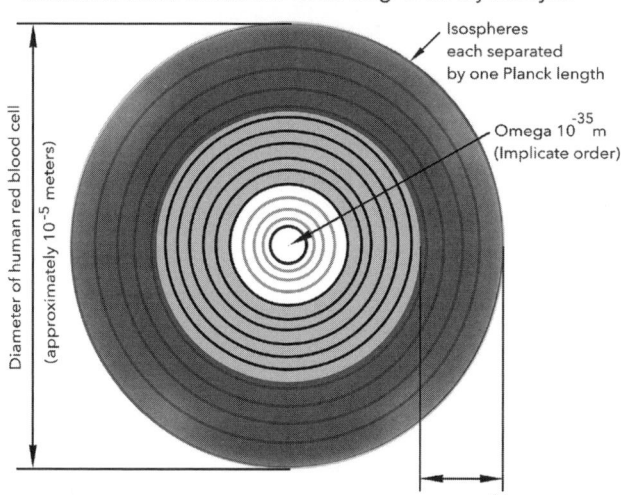

How 7 billion unique human frequencies can be accommodated within the outer ring of an erythrocyte.

Isospheres each separated by one Planck length

Omega 10^{-35} m (Implicate order)

Diameter of human red blood cell (approximately 10^{-5} meters)

Fig. 11.8. Isospheric capacity of a single erythrocyte.

7 billion human isospheres separated by 1 Planck length in a 10^{-26} meter range. (not to scale)

Dimensional Size Considerations

In terms of dimensional analysis (size considerations in the spatial dimensions) Jung's view is congruent with contemporary theories in the physics of consciousness such as Roger Penrose and Stuart Hameroff's "Orch-OR" hypothesis, that insists that consciousness at the granular level is being generated in an orchestrated process within human microtubule cavities at the Plank length (10^{-35} m) dimensional scale (at the far left in fig. 11.9). Instincts ("psychoidal subroutines") may, by contrast, be operational at less energetic levels at longer wavelengths, possibly in the scalar range of neurons (10^{-4} m) in the infrared band of wavelengths. Recent discoveries in radio astrophysics have detected enormous wavelengths the size of galaxies that may imply forms of communication among mega formations of galactic psychoids throughout the universe.

Assuming each unique frequency would match its radially unique isosphere, separated by only one Planck length, figure 11.8 suggests how 7 billion unique isospheres, each of quantum discrete frequency, might be nested within the geometry of a typical human red blood cell (in the image, a multiple of 7 billion times the Planck length of approximately 10^{-35} m results in an estimated shell thickness of 10^{-26} m). This model supports the feasibility that each living human being might have a unique holospheric frequency, detectable by other human blood cells via the implicate order about which each is centered, and thus provides a possible mechanism for communication, via the mechanisms of resonance, nonlocality, and superposition in the frequency domain of the implicate order.

Energy Spectrum of the Psyche

10^{-35} meter ←――――――――――――――――――――→ 10^{26} meter

| Region of Planck length Holospheres | | Human Dimensions | | Region of Diameter of the Universe |

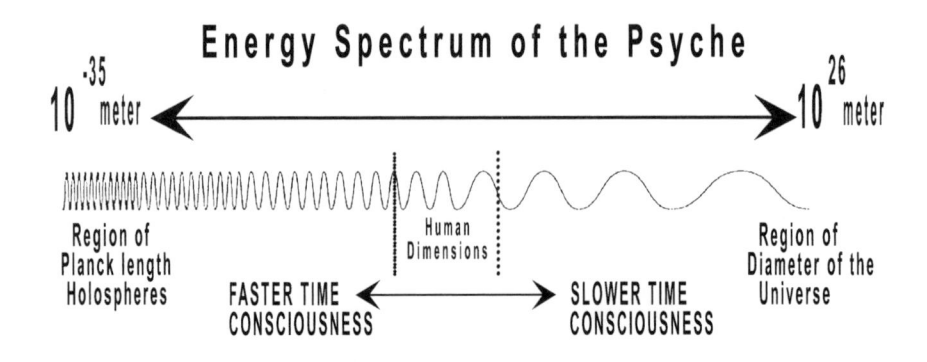

FASTER TIME CONSCIOUSNESS ←――――→ SLOWER TIME CONSCIOUSNESS

Fig. 11.9. The energy spectrum of the psyche.

The Fourier Transform:
Linking Spacetime and the Hidden Dimensions

If consciousness is a phenomenon of dynamic fields of resonating electromagnetic holoflux energy, how then, might consciousness span multiple dimensions even beyond space and time? How might consciousness store, move, and perceive information among the hidden dimensions that are beyond our space and time dimensions?

As we have seen, Carl Pribram's lifetime quest was to understand how and where memory is stored. After years of surgical brain research he found that the location for memory storage could not be found anywhere in the tissue of the brain. In the mid-1960s Pribram found an answer in the mathematical relationship of the Fourier transform, an equation showing how information flowing in spacetime can be projected into dimensions beyond space and time (and vice versa). This is a subtle yet powerful concept that has guided computer and communication engineers to the successful development of Fourier transform holography, which has revolutionized science, medicine, and even metaphysical thinking. The Fourier transform suggests how information encoded in spacetime is linked directly with the other seven dimensions predicted in the superstring M-theory, via the dynamics of time.

As previously discussed in chapter 9, the mathematician Jean Baptiste Fourier, working at the time as Napoleon's governor of Egypt, managed to derive a mathematical equation to model, in terms of pure frequency, the flow of heat energy in time and space. While the work of Euler had earlier proven the initial link between the frequency and spacetime domains, it was Fourier, who, during his experimental investigation of heat flow, was able to derive a mathematical operation of integral calculus that expressed accurately the energy transformations between a time and pure frequency. While this allowed the more accurate casting of large iron cannons, it also provided the missing link between the time and space dimensions and pure frequencies, which greatly assisted in the mathematical calculations of heat flow required to cast a perfect cannon barrel.

This discovery has great significance for the twenty-first century study of consciousness, as the Fourier transform can be understood as the link that bridges spacetime and dimensions beyond spacetime via the frequency domain. It is the frequency domain that is common to all dimensions in

$$f(t) = a_0 + \sum_{n=1}^{\infty} \left(a_n \cos \frac{n\pi t}{L} + b_n \sin \frac{n\pi t}{L} \right)$$

Fig. 11.10. The Fourier series.
Image from Chen, *Introduction to Plasma Physics and Controlled Fusion*, 1.

the metaverse, in particular those beyond spacetime. These Fourier transforms are themselves derived from an underlying series of alternate pure sine and pure cosine waves (fig. 11.10). Sine and cosine signals are the basis for all analog communication devices.

A century after Fourier's death, Norbert Wiener made use of Fourier's transform to model and analyze brain waves, and he was able to detect a range of frequencies, centered within different spatial locations on the cortex that exhibited awareness of each other. Specific frequencies were found to be attracting one another toward an intermediate frequency, thus exhibiting resonance or "self-tuning" within a narrow range of the frequency domain. This discovery led Wiener to conjecture that the infrared band of electromagnetic flux may be the loci of "self-organizing systems."[17]

> We thus see that a nonlinear interaction causing the attraction of frequency can generate a *self-organizing system*, as it does in the case of the brain waves we have discussed [. . .] This possibility of self-organization is by no means limited to the very low frequency of these two phenomena. Consider self-organizing systems at the frequency level, say, of infrared light.[18]

Wiener goes on to discuss the possibilities of communication through electromagnetic frequencies in biology, where he focuses upon the problems of communication at molecular and primitive cellular levels, specifically on the problem of how substances produce cancer by reproducing themselves to mimic preexisting normal local cells. Molecules do not simply pass notes to one another, and they do not have eyes, so how do they perceive and how do they communicate? Wiener conjectures:

The usual explanation given is that one molecule of these substances acts as a template according to which the constituent's smaller molecules lay themselves down and unite into a similar macromolecule. However, an entirely possible way of describing such forces is that the active bearer of the specificity of a molecule may lie *in the frequency pattern of its molecular radiation*, an important part of which may lie *in infrared electromagnetic frequencies* or even lower. It is quite possible that this phenomenon may be regarded as a sort of attractive interaction of frequency.[19]

Similarly, Jung observed the direct relationship of psyche and mind to matter, noting that they are "two different aspects of one and the same thing."

Since psyche and matter are contained in one and the same world, and moreover are in continuous contact with one another and ultimately rest on irrepresentable, transcendental factors, it is not only possible, but also fairly probable, even, that psyche and matter are two different aspects of one and the same thing.[20]

Like Jung and Bohm, electrical engineers map the world into two domains, which they term the *complex-frequency domain* and the *time domain*. It is our normally perceived spacetime that electrical engineers identify as the *time domain*, but it is in the *frequency domain* that they see the operation of communication signals.[21] These two orders of electrical engineering theory can be seen to reflect David Bohm's explicate and implicate orders. Bohm's explicate order can be associated with the time dimension through which signals are transmitted and received sequentially in time, while Bohm's implicate order can be identified with the electrical frequency dimension as a timeless, spaceless domain of pure frequency-phase information.

All of these assumptions support the idea that it is the complex-frequency domain that links all of the dimensions via pure frequency patterned information. One is reminded of the most famous symbols of Hinduism, the mantra associated with cosmic sound that is taught in the Vedas: Ôm. In Sikhism, it is called Onkar, which means "the One Ôm," of which it is said in a text from 930 CE:

*Onkar ("the Primal Sound") created Brahma, Onkar fashioned
 the consciousness,
From Onkar came mountains and ages, Onkar produced the
 Vedas,
By the grace of Onkar, people were saved through the divine
 word,
By the grace of Onkar, they were liberated through the teachings
 of the Guru.**

It appears that mystics have long known what physicists over the past century have slowly learned through the mathematics of quantum mechanics and electrical engineering!

Each Soul as a Multidimensional Hologram

Superimposing Carl Jung's view of the soul with the Pribram-Bohm holoflux hypothesis, the soul can be seen as a unique pattern of energy, a multidimensional hologram whose axis runs through each of the eleven dimensions hypothesized in M-theory.

The congruence of the point of intersection of these many dimensions, at the very center of the soul, has been observed in the work of several notable individuals:

- It has been called Omega in the work of Teilhard de Chardin, a point that he also calls, the "Christic," at the very center of each soul's being.
- It has been called "the still point" by the poet T. S. Eliot, who, as noted above, indicates in his poem "Burnt Norton" in *The Four Quartets* that "there the dance is," going on to say that it is in a domain that is spaceless and timeless, and yet, except for the still point, there would be no dance, and "there is only the dance."
- It has been called an "actual entity" (or "actual occasion"), by the mathematician turned philosopher Alfred North Whitehead, for whom actual entities exist as the primary foundational elements of

---

*From the *Adi Granth* by Ramakali Dakkhani.

reality, the ultimately existing facts of the world, underlying all of reality, both in time and in the transcendent dimensions beyond time.
- It has been called the "soul" by Jung, the center of which is the *interface between* the psyche (the totality of all consciousness spanning every dimension) and the individual ego's soul (the isolated "little psyche").
- In the context of Bohm's sub-quantum theory of consciousness, the soul can be visualized as *the link* between spacetime and the implicate order at the very center of every conscious being. More specifically this link, where the action of the Fourier continuously occurs, operates at the locus of an infinitely thin spherical shell at the boundary (the event horizon or Plank holosphere) between spacetime and the other dimensions, all of which are transcendent with spacetime.

Thus each soul manifests as a multidimensional hologram, distinguished by its own unique holoflux signature. The soul can be visualized as a multidimensional holoflux of dynamic energy operating not only within spacetime, but inextricably interconnected with all of the atemporal dimensions. Its absolute position in the multidimensional metaverse can be identified through specification of all of the dimensional "point" coordinates taken together, and the trajectory function "f" (dynamic path) of a psychoid (soul or actual entity) can be mathematically expressed in the following function (assuming the eleven dimensions indicated by string theory):

$f(x, y, z,$ time, dimension 5, dimension 6, dimension 7, dimension 8, dimension 9, dimension 10, dimension 11$)$.

Note that a typical function f plots x, y, z and sometimes t. However, the absolute location of the actual entity in the multidimensional universe is the intersection of every dimension at the one unique point indicated by the function. We are indeed unique in the metaverse!

Part 5

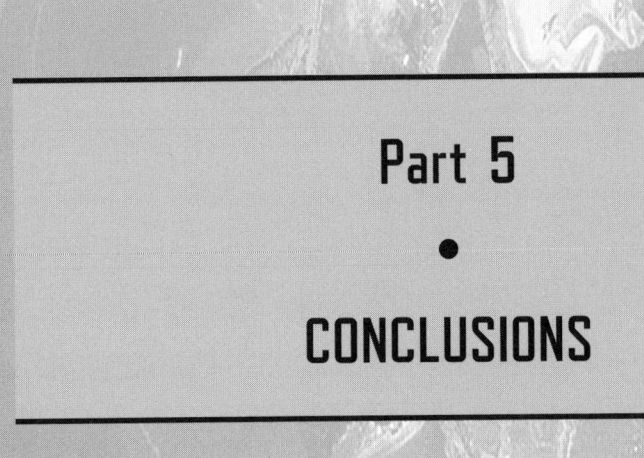

•

CONCLUSIONS

12

Exploring the Hidden Dimensions of the Mind

Throughout our lifetimes, each one of us views, interprets, and understands the universe from a truly unique vantage point in space and time. Billions upon billions of moments of experience, thoughts, memories, and sensations accrue in our personal lifetime, resulting in an increasingly unique perspective and the patterns of thought that make each one of us unique. At birth, our small seemingly separate awareness awakens with the merging of two parental streams of consciousness, forming a nexus of awareness that is the new human psyche. This infant awareness opens its inner eye, rising up from the depths of consciousness, becoming a small new island of awareness surrounded by the dark depths of consciousness. Seeded with the inheritance not only from two parents, but from the experiences of their entire ancestral history, this new, unique awareness begins to accrue experience. New experiences affect the brain as it grows. A new, unique personality, whether labeled purusha, soul, or ego, this unique mind continuously changes as it accrues a lifetime of new experiences. How does the human brain structure this new human mind?

One Brain: Two Hemispherical Minds

A clue to the structure of the psyche operating as a human mind within a human brain can be found in observations collected from decades of split-brain research. In the architecture of the psyche, each human mind is

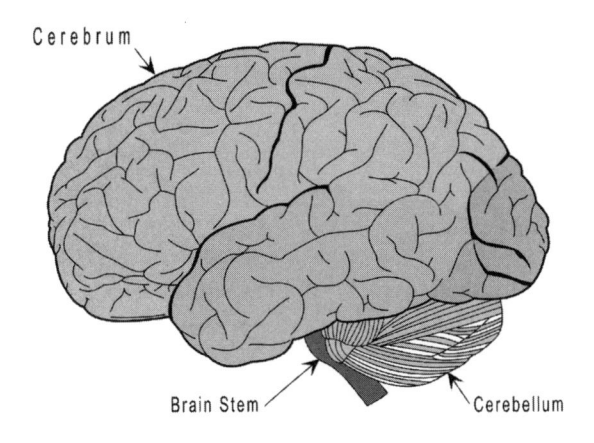

Fig. 12.1. Cerebrum, cerebellum, and brain stem.

a dynamic system generated and sustained by energy flowing through two sides of what is called the cerebrum (fig. 12.1).

Comprising 85 percent of total brain weight, the cerebrum is the largest part of the brain and sits on the top of the Brain Stem. It is also the most recently developed region in the brain's evolutionary history.

The cerebrum is split into two physically separate halves: the left and right cerebral hemispheres (fig. 12.2). The only physical connection between the two is a mass of 300 million axons (nerve fibers) called the *corpus callosum*. After instances in which medical treatment (as in the case of grand mal epileptic seizures) necessitates completely severing this thick white mass of nerve fibers, the frequent postoperative result is that patients have often been observed to exhibit behavior as if they were two entirely distinct (and not always compatible) personalities.

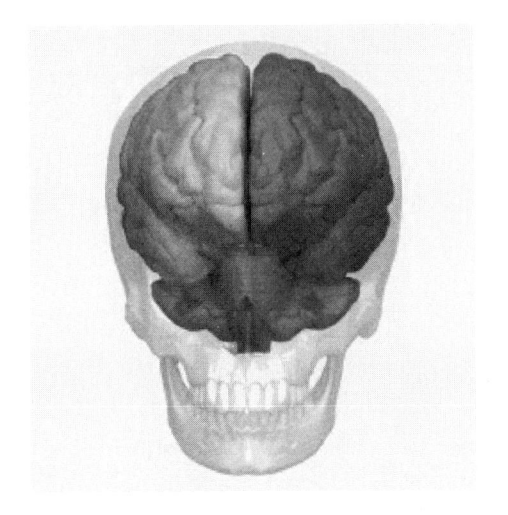

Fig. 12.2. Split brain: view of cerebrum divide. Image by Database Center for Life Science (DBCLS) and BodyParts3D.

This phenomenon suggests that what we typically consider our single conscious self is likely made up of two minds that usually, but not always, work together as one, similar to the way two adults function in a marriage. Years of split-brain research suggest that each human being may possess two distinct centers of cognition, or "sub-personalities," which typically function together so seamlessly that they present as a single, unified mind. If a contemporary individual's mind is actually the product of a dynamic relationship between these two sub-personalities, this understanding could have profound implications for the treatment of personality disorders.

A therapist who recognizes that a patient might consist of two differentiated personalities experiencing internal conflict might approach the analysis and treatment of symptoms very differently than if the patient were assumed to have a single, unified personality. This perspective isn't limited to scientific research; it resonates with cultural insights as well. For instance, Rastafarians often use the phrase "I and I" to express the idea that each person embodies two entities—an "I" and another "I"—a duality that modern culture generally overlooks (other than in individuals diagnosed with schizophrenia).

Using the shorthand "LH" and "RH" for "Left Hemisphere" and "Right Hemisphere" personalities, we can observe that in a healthy, well-integrated individual, the LH and RH work cooperatively, sharing common goals and values. However, in individuals with personality disorders, the LH and RH may have conflicting goals and values that require resolution. Problems also arise when one hemisphere dominates the other, suppressing its counterpart.

This understanding complicates the traditional view of two-person, or dyadic, relationships. Previously, it was assumed that such relationships (e.g., between a wife and husband, mother and child, or siblings) involved only two personalities. Consequently, therapists have classically approached these relationships with the goal of addressing the dynamics between two distinct personalities. However, split-brain research suggests that each individual actually brings two distinct personalities into the relationship, leading to a total of four interacting personalities. This means that any one of the four sub-personalities could potentially be in conflict with any of the other three. Thus, the dynamics between two people are far more complex than the formerly assumed interaction between two centers of personality, making it a far more intricate, four-way relationship.

Assume, for example, that a newborn's brain carries holonomic imprints from both parents—perhaps the left hemisphere bears traces of the father's personality, while the right hemisphere carries those of the mother. As the child grows, life experiences are layered onto these initial configurations and integrated through superposition. Thus, when two humans interact, it is not just a relatively simple meeting between two individuals, but rather a complex interaction among four mind centers or "hemispherical personalities." The six simultaneous personality relationships are shown in figure 12.3.

While these sub-personalities often align harmoniously, this isn't always the case. Conflicts or incompatibilities between any of the sub-personalities can lead to personality disorders, relationship issues, and challenges within the overall relationship. Wider acceptance of the split-brain multiple personality architecture would have an enormous impact on understanding and treating issues such as schizophrenia, transgenderism, and multiple personality disorder.

In general, the left hemispherical side of the brain has been observed to be responsible for language and speech; because this is how we present ourselves to the world at large, it has been called the "dominant" hemisphere.

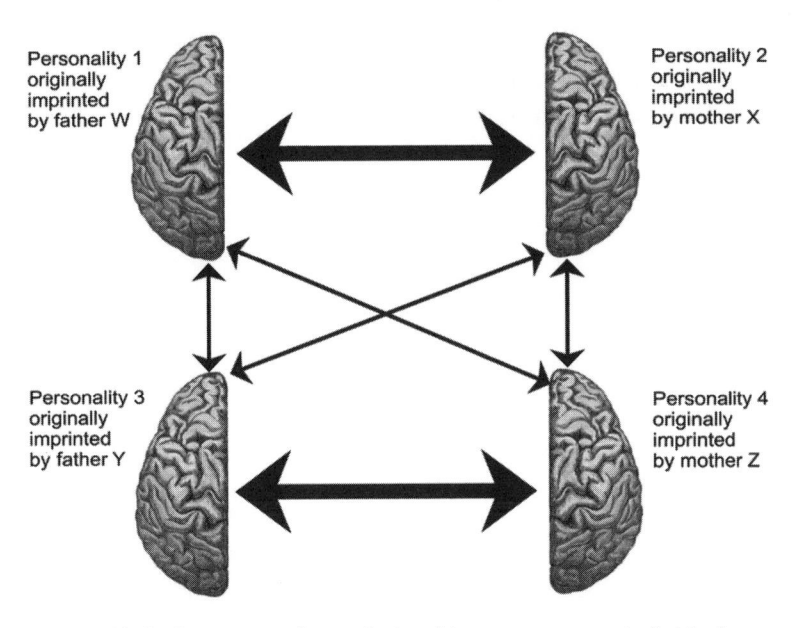

Fig. 12.3. Six personality relationships among two individuals.

By contrast, the right hemisphere plays a bigger part in interpreting visual information and spatial processing.

The following table offers some of the observed differences.[1]

Differences between Left Cerebral Mind and Right Cerebral Mind

Left Hemisphere: Rational	Right Hemisphere: Intuitive
Linear thinking (sequential)	Parallel thinking (holonomic)
Looks for differences (differentiates)	Looks for similarities (integrates)
Searches for understanding through reasoning	Searches for understanding through insight and intuition
Proclivity for critical thinking	Proclivity for feeling emotions
Fascination with numbers	Fascination with colors and sounds
Is tightly planned and structured	Is fluid and spontaneous
Parses and analyses information for individual details	Overlays information to obtain a wider patterns
Solves problems by logically and sequentially looking at small parts	Solves problems intuitively, searching for large connecting patterns
Communication through talking and writing	Communication through music, drawing, painting, and touch
Tightly restrains expression of emotional feelings	Freely expresses emotional feelings
Memory adapted for recollection of symbols and spoken and written messages	Memory adapted for recollection of music, color patterns, static and kinetic body positions and sequences
Responds best to audible verbal instructions	Responds best to visual instructions

Your "mind" can thus be seen as a loosely integrated combination of two separate but superimposed centers of cognitive function and awareness, not always coequal. It is as if two psyches are active simultaneously within a single cranium. These two architecturally separate conscious entities evolve over each individual's lifetime from the initial imprinting passed on from the psyches of mother and father during the moments of procreation. It is thought that the left-hemisphere's psyche holographically inherits the complex of one's father (and his own family predecessors) while the right-hemisphere's psyche is imprinted with the complex from one's mother (and her line of family ancestors). From that initial imprinting, one's life experiences begin to overlay and supplement the initially inherited configurations of the psyche. As time progresses, one acquires layers of new experience through interactions with the world, and the patterned programming of each hemisphere's actual entity grows even more unique.

It has been widely observed among postoperative individuals who have had the corpus callosum severed, that the two quasi-independent psyches are not always compatible, that is, they do not always seem to agree with one another. In some instances, they have been observed to act aggressively at odds with one another.

In most individuals, whether or not the corpus callosum has been severed, one of the two hemispheres usually dominates the other. In a similar way we say a person can be seen to be "right-handed" or "left-handed." One can surmise that the pressures of social conformity in patriarchal societies may further dominance of the left hemisphere's analytical, verbal skills over the right hemisphere's integrative, intuitive strengths.

Modern educational methodologies have arisen to balance the inherent left-right hemisphere attributes through the equal development of both cognitive and creative attributes. The ensuing integration and ensuing harmony is seen to be highly beneficial, resulting in the healthy operation of the composite mind and a healthy psyche. Lack of such integration can lead to dysfunctional operation of the "two minds" within the seemingly single individual.

This suggests the importance of a balanced education early in life, one that offers language skills, mathematical abilities, reasoning and analytical skills to develop the left-hemispherical personality, while, on the other hand, offering training in the visual arts, music, literature, dancing, and social-emotional interaction skills for the right hemisphere. Rudolph Steiner actually developed such a balanced system that has become the foundation of the Waldorf school system now found in many countries. Unfortunately, modern education now stresses logic, reason, and language at the expense of the arts and social skills, an imbalance that is likely reflected in many of the current problems within societies throughout the world.

The Third Mind (beyond the Brain)

In addition to the evidence from studies of patient behavior after corpus callosum severance, both religious tradition and contemplative introspection offer additional supporting evidence that there is a third center of consciousness or "mind" that can be associated with the human psyche, traditionally called *soul* in the Christian culture, *purusha* in Hindu culture, and *Tao* in Chinese culture.

The renowned mystic G. I. Gurdjieff strongly supported the idea of this third mind. Gurdjieff wrote extensively about humans having multiple mind centers (many "small I's") and characterized humans as "three-brained beings." In 1927, Gurdjieff completed his 1,248-page allegorical novel *Beelzebub's Tales to His Grandson: All and Everything*, that includes a fictional account of the firsthand experience of an interstellar traveler visiting the Earth. Early in the book it is noted that humans on this planet have evolved as "three-brained beings." In relating the history of humankind, Gurdjieff covers the essential elements of his system of conscious development and describes how the vast majority of beings on planet Earth live their entire lives in a state of hypnotic "waking sleep." Nevertheless, he tells us, it is possible to transcend to a higher state of consciousness awareness and achieve full human potential through quieting all of one's "little 'I's" and thus allowing the third center of consciousness to rise up into fuller awareness.

With sufficient training, this third mind begins to actively function in the foreground with active links to the other dimensions of the metaverse. This "third-brain" or transcendental mind, operating beyond space and time, is the actual substrate or foundation, the Self as described by Jung, resonating with the two human spacetime mind centers, the left cerebral hemispherical mind and the right cerebral hemispherical mind. In India, this Self is known as the *purusha* and thought to be the root of our true separate consciousness, emanating directly from Brahma that which is "unchanging, eternal, and pure," and transcending space and time. Each purusha is a single seemingly separate actual entity, a diffraction of the One Whole, the Brahma.

One of the goals of the psychotherapeutic process pioneered by Jung is to assist the individual ego to open up to this larger Self, to enable a healthy interrelationship to be established between the normally seemingly isolated, individual ego of the human individual and this Self that bridges the metaverse of dimensions (and the myriad of psychoidal archetypes within these dimensions) in its entirety. Jung termed such a healthy connection between ego and archetypes an "individuation," a state in which the ego operates in a balanced, healthy, whole manner as an integral part of an ocean of collective psychoids that make up the greater unitary metaverse.

Many contemplative traditions teach individuals how to practice mental exercises that can lead to the development of the capacity to silence normal mental activities in order to open the portal to this wider dimensional metaverse. During the period of sustained contemplative silence of the two cerebral "laptop-minds," a previously unperceived panorama arises into awareness, a scintillating holonomic flow of holoflux that some have called "cosmic consciousness" (Bucke) or "one taste" (Wilber). The new awareness is not merely visual, but a combination of sensations perceived through direct contact with multiple dimensions beyond normal human sensory systems. At some moment during meditation, one experiences a "rupture of plane" (McKenna) as the formerly dualistic cognitive identity fades into the background, replaced with a deep and immediate sense of having joined a cosmic stream of flowing energies.

This flow of holoflux energy is not just a movement in time and space, but is a holonomic flow within and between the many dimensions beyond spacetime, including the deepest source dimensions that are the foundation of all, as discussed here in this dialog between David Bohm and Jiddhu Krishnamurti:

BOHM: Would you say energy is a kind of movement?

KRISHNAMURTI: No, it is energy. The moment it is a movement it goes off into this field of thought.

BOHM. We have to clarify this notion of energy. I have also looked up this word. You see, it is based on the notion of work; energy means, "to work within."

KRISHNAMURTI: Work within, yes.

BOHM: But now you say there is an energy which works, but *no movement*.

KRISHNAMURTI: Yes. I was thinking about this yesterday—not thinking— I realized *the source* is there, uncontaminated, *non-movement*, untouched by thought, it is there. From that these two are born. Why are they born at all?

BOHM: One was necessary for survival.

KRISHNAMURTI: . . . In survival this—in its totality, in its wholeness—has been denied, or put aside. What I am trying to get at is this, Sir. I want to find out, as a human being living in this world with all the chaos and suffering, *can the human mind touch that source in which the two divisions don't exist?* —and because it has touched this source, which has no divisions, it can operate without the sense of division.[2]

Thus we are not the "person" we think we are; in fact, we are much, much more. We are the universe looking out from a particular perspective into time and space, observing and interacting with itself.

Opening up to the Self and the Metaverse

So how does one begin to open up one's own portal of consciousness to the wider Self and thus to the metaverse? There are many techniques that have been discovered down through the ages in most, if not all cultures, passed down as skillful means in religious and shamanic traditions. Some traditions have identified the ingestion of plant entheogens as a beneficial approach for opening one's consciousness to the initial experience. Other methods stress the efficacy of physical exercises such as yoga, tai chi chuan, and even dervish dancing for catalyzing entry into the silence. The primary emphasis in all of these approaches is the focus upon learning to enter a state of cognitive silence. Only through the silencing of the normal mind (called "the monkey mind" in some traditions) can the experience of the deeper pervasive silence arise.

To explore the hidden dimensions of consciousness, one must, through practice, learn to quiet the normal activity of the brain, sometimes referred to in yoga as "quieting the monkey mind." This requires "letting go" of one's normal cognitive flow (i.e., random thoughts, internal verbal dialog, memory sequences) while simultaneously reducing sensory awareness of warmth, cold, discomfort, and external sounds. Texts of classical yoga recommend meditating in silent caves or the desert, while my own practice for many years has been to meditate in a dark room at night while wearing earplugs. My mentor John Lilly practiced meditation for long periods floating in body-temperature salt water in a lightproof, soundproof chamber of his own design, often under the influence of ketamine or LSD.

Eventually, with consistent practice, one can significantly diminish the flow of thoughts and distraction of external sensations. At some point, emerging from the silence, a new mode (Ken Wilber calls it a new "flavor") of consciousness begins to enter awareness, an unusual but unmistakable sense of joining a living network of consciousness, perhaps even more than one distinct network. There is a pronounced sensation of actually touching something and being touched by something. One's field of awareness opens up to something that is much bigger, wider, and deeper than normal sensory input, and though first one may sense a fear of this unknown vastness, the fear is soon replaced with a sense of belonging, of being welcomed. In some traditions this is termed the experience of the primordial consciousness, in Sanskrit the *dharmadhātu*, that underlies the entire fabric of reality, the doctrine of the Void, or *śūnyatā*, that is the goal of contemplative practices in Tibetan Dzogchen, Chan, and other Buddhist traditions.

This experience of the Void is here described by the psychiatrist Stanislav Grof:

> The Void, primordial Emptiness and Nothingness is consciousness itself. The Void has a paradoxical nature; it is a vacuum, because it is devoid of any concrete forms, but it is also a plenum, since it seems to contain all of creation in a potential form. According to Ervin Laszlo, the Void is a subquantum field which is the source of all creation and in which everything that happens remains holographically recorded. Laszlo equates this field with the concept of quantum vacuum that has emerged from modern physics.[3]

Bohm, from the viewpoint of quantum physics, also clearly stated that the void is not simply an empty vacuum but a plenum: "Space is not empty. It is full, a plenum as opposed to a vacuum, and is the ground for the existence of everything, including ourselves. The universe is not separate from this cosmic sea of energy."[4]

In order to reach this experience of the void, our memories, thoughts, and external sensations need to be ignored and allowed to fade away. My own first experience of opening to these hidden dimensions occurred (described at length in chapter 2) in a quiet inner room of my fifth-floor walkup apartment, while doing a hatha yoga pose in the silent darkness late at night. Suddenly from out of the silence, a loud, pure, high-pitched whistling sound arose into my awareness. Over the ensuing years my perception of this sound has grown richer, with increasingly complex "subtle sounds" and various feelings of awareness and intuitions, inexpressible in terms of space and time and words.

I later discovered that this subtle awareness is not as rare as I had thought. Medical research has determined that 15 percent of the adult population experiences what is called *tinnitus* at some time in their lives. It is thought to be to be a widespread issue that affects millions of people globally. Modern medicine treats tinnitus as a disease, as it is often disconcerting and anxiety-provoking to those who begin to hear the tones, sometimes quite loud. But what if tinnitus is not a disease? Could it simply be the beginning of an opening to the hidden dimensions of the void? What if people, instead of running from the phenomenon, choose instead to begin exploring, even cultivating these sounds, rather than trying to suppress them in fear after being told that they are somehow a physical disease? Imagine the first sea creatures responding with fear at the unknown new sensory input of the sensation of light as their rudimentary eyes began to open awareness to the sun. My own conjecture, as a computer and electrical engineer, is that this substrate of sound frequencies that I hear when my brain quiets down, is an opening to a sort of cosmic machine language, an inner cosmic network operating at extremely high frequencies (at very small, perhaps sub-quantum dimensions). This might explain my own experiences on ayahuasca of sensing a link or portal opening up to a wider awareness of the universe, one that seemed to examine my body on many levels and somehow then tuned up various systems within me.

Conclusions and Recommendations

In our search for guidance to help us understand and better navigate the profound challenges we face, both as a planet and as individuals, we have turned in this book to the rich maps of consciousness developed by Carl Jung, Teilhard de Chardin, and Gustav Fechner. Each of their cosmologies locates the consciousness of an individual human within a wide spectrum of conscious entities that operate outside of ordinary human awareness in vast, largely uncharted dimensions. Much of their insight and detailed theories were developed through direct personal experiences of consciousness. Their common methodology was to engage regularly in direct contemplative observation of the movement of their own minds, a technique pioneered by Wilhelm Wunt (1832–1920) that he termed *introspection*. Unfortunately, the practice of introspection was eventually dismissed and even denigrated by modern scientific materialists who have grown ever more entrenched in their myopic insistence that the *only* data that can be accepted *must be* of objective, repeatable, recordable behavior, and definitely not obtained through "observing the mind with the mind itself." Thus, the rich integrated maps developed by Jung, Teilhard, and Fechner, potentially answering many questions about consciousness and pointing the way toward further research, were often disregarded and frequently excluded from "serious" scientific discussion.

But recently the theories of Jung, Teilhard, and Fechner have been gaining new recognition and respect due to advances in particle physics and mathematics. String theory and high-energy particle experiments now support the existence of at least seven dimensions beyond spacetime, and developments in artificial intelligence and neuronal networking highlight the possibility of vast, interconnected networks of conscious agents (psychoids) extending weblike throughout multiple dimensions of reality. The writings of Jung, Teilhard, and Fechner all suggest that the human brain and mind do possess the capacity to tune into and to join these other levels or networks of consciousness. Once a link with other modalities of consciousness has been established, whether through meditation, entheogens, prayer, or technological advancements (e.g., so-called "God helmets," or brain-implanted emf devices), the connection, tenuous at first, can be strengthened through regular practice, eventually becoming a channel for communication with other psychic entities for the potential reception of knowledge, guidance, and insight, from other sources of consciousness.

The idea that consciousness may manifest in frequency vibrations that can enable links to hidden conscious entities is not a new one. The

Swedish philosopher Emanuel Swedenborg (1688–1772) was a trained scientists who wrote extensively on his regular communication with conscious entities that are beyond normal human perception. In his 1718 dissertation, "On the Mechanism of the Operation of the Soul and the Body," Swedenborg developed a theory of vibrations or subtle movements—what he referred to as "tremulations"—as a way of understanding how thoughts, sensations, and psychic influences could affect the physical body. In his writings, particularly those that blend his scientific knowledge with mystical theology, he often referred to these vibrational phenomena as a medium for the transfer of spiritual energies or divine influence on the material world.

There are accounts from Emanuel Swedenborg's contemporaries, including his servants, that they claimed to have witnessed him conversing with invisible spirits. These stories gained attention during the later part of Swedenborg's life, when he became deeply involved in his internal experiences and began to write extensively on the subject of his interactions with the spiritual world. Swedenborg himself claimed that from around 1744 onward, he was able to converse directly with angels, spirits, and even deceased individuals from the afterlife.

Accordingly, I encourage you to join with the many explorers of consciousness such as Swedenborg, Jung, Fechner, Teilhard de Chardin, and many more in establishing connections between humanity and denizens of the hidden dimensions by forging connections that reach out into the metaverse through regular contemplative practice. Learn to tune your mind as an instrument that can become capable of making that initial contact with new dimensions. At a minimum, your ability to relax into a sustained silence, free from spontaneous mental processes, will give your cerebral cortex a rest and quiet the activity of your two cerebral hemispheres, during which your cortex will no longer be acting as a Faraday cage to shield your inner brain regions from external signals. The psyche then has the possibility of opening out into a deeper awareness of new dimensions beyond the isolated self (fig. 12.4.).

With effort and intent, one can establish links to rich networks of consciousness (Teilhard's noospheres) that can enhance one's ability to receive intuitive knowledge, while also maintaining a calm, tranquil, centered state of being in the midst of a world filled with anxiety, change, and

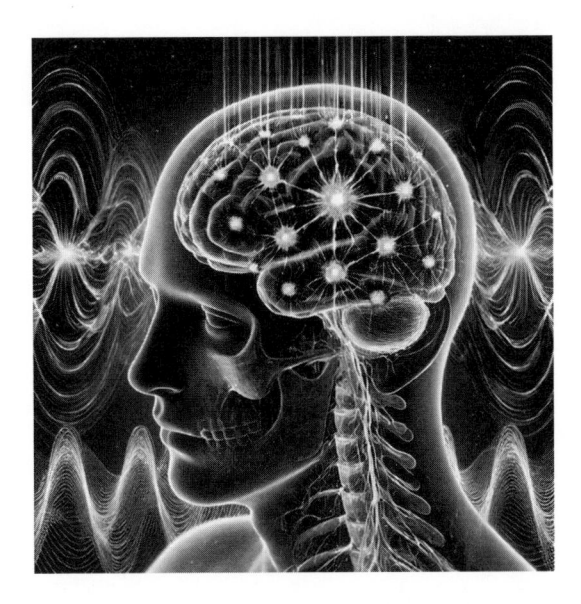

Fig. 12.4. Neurons resonating with hidden dimensions of consciousness.

confusion. Such a regular practice of meditation and contemplation can enrich your being while simultaneously offering one of the few real ways that an individual can open new avenues of possibility to join in solving the many problems that now engulf our civilization on the planet Earth.

I would like to close with a quote from Gustav Theodor Fechner:

Beside our consciousness there is still more consciousness, that over and above all individual consciousness there is a broader and higher consciousness with broader and higher content, a consciousness which on the side upon which it excels and surpasses our consciousness represents the outward world by which our consciousness is determined, and ties together all individual consciousness by common situations and effective relationships, the highest unity of which is found in the last knot, love.[5]

NOTES

INTRODUCTION

1. Zeilik and Gregory, *Introductory Astronomy & Astrophysics* (4th ed.).
2. Minkowski. "The Fundamental Equations for Electromagnetic Processes in Moving Bodies," 1–69.
3. Hoffman, "Consciousness, Mysteries Beyond Spacetime" 40.45 (video).
4. Bohm and Weber, "The Physicist and the Mystic—Is a Dialogue between Them Possible?," 27.

CHAPTER 1. THE HIDDEN DIMENSIONS OF CONSCIOUSNESS

1. Carr, *Universe or Multiverse?*, 10.
2. Sutter, "How the Universe Could Possibly Have More Dimensions."
3. Sutter, "How the Universe Could Possibly Have More Dimensions."
4. Steven Hawking in foreword, May and Israelian, *STARMUS: 50 Years of Man in Space*, xvi.

CHAPTER 2. UNIVERSE OR METAVERSE?

1. Wallace, *The Art of Transforming the Mind*, 112.
2. Irenaeus, *Adversus Haereses* [Against Heresies] II, 75.
3. Schroll, "Understanding Bohm's Holoflux," 132.
4. Steiner, *Knowledge of the Higher Worlds and Its Attainment*, 1.
5. Steiner, *The Archangel Michael: His Mission and Ours*, 73.
6. Lilly, *The Scientist: A Metaphysical Autobiography*, 128.

CHAPTER 3. HOW I CAME TO EXPLORE THE HIDDEN DIMENSIONS

1. Crabtree, *From Mesmer to Freud: Magnetic Sleep and the Roots of Psychological Healing*.
2. Eliade, *The Road to the Center*, 445.
3. Lilly, *The Scientist: A Metaphysical Autobiography*, 128.
4. Blofeld, *Tantric Mysticism of Tibet: A Practical Guide to the Theory, Purpose, and Techniques of Tantric Meditation*; Churton, *Aleister Crowley: The Biography*.
5. Blofeld, *The Wheel of Life*, 151.
6. Blofeld, *The Wheel of Life*, 153.
7. Churton, *Aleister Crowley: The Biography*, 108.
8. Albert, Metzner, and Leary, *The Psychedelic Experience*, 1.
9. Ouspensky, *In Search of the Miraculous*, 3.
10. Wood and Brunton, *Practical Yoga, Ancient and Modern*: *Being a New, Independent Translation of Patanjali's Yoga Aphorisms*.
11. Taimni, *The Science of Yoga*.
12. Whicher, *The Integrity of the Yoga Darsana: A Reconsideration of Classical Yoga*, 42.

CHAPTER 4. FECHNER'S PSYCHOPHYSICS OF THE HUMAN SOUL

1. James, Introduction to Fechner, *The Little Book of Life after Death*, vii.
2. Fechner, *Concerning Matters of Heaven and the World to Come*, 50.
3. James, introduction, Fechner, *The Little Book of Life after Death*, xvii.
4. James, introduction, Fechner, *The Little Book of Life after Death*, xii.
5. Fechner, *The Little Book of Life after Death*, 65.
6. James, introduction, Fechner, *The Little Book of Life after Death*, xiii.
7. James, introduction, Fechner, *The Little Book of Life after Death*, xii.
8. Fechner, *The Little Book of Life after Death*, 53–54.
9. Kahneman, *Thinking, Fast and Slow*, 72.
10. Heidelberger, *Nature from Within: Gustav Theodor Fechner and His Psychophysical Worldview*, 127.
11. James, *The Varieties of Religious Experience*, 441–42.
12. Fox, "Gustav Fechner: The Man Who Introduced Soul to Science," 4.

CHAPTER 5. JUNG'S SPECTRUM OF THE PSYCHE

1. Jung, *Collected Works* 6:797 [emphasis added].
2. Jung, *The Red Book: A Reader's Edition*, 5.
3. Jung, "Spirit and Life," in *The Structure and Dynamics of the Psyche*, 337.
4. Jung, "Spirit and Life," 215.
5. Teilhard, "Human Energy," 147.

6. Jung, "On the Nature of the Psyche," in *The Structure and Dynamics of the Psyche*, 215.
7. Jung, "Spirit and Life," 320.
8. Jung, "Spirit and Life," 335.
9. Jung and Pauli, *The Interpretation of Nature and the Psyche*.
10. Jung, "On the Nature of the Psyche," 185.
11. Jung, "On the Nature of the Psyche," 210.
12. Jung, *Collected Works* 6:797 [emphasis added].
13. Addison, *Jung's Psychoid Concept Contextualized*.
14. Jung, *Collected Works* 13:350.
15. Aziz, *C. G. Jung's Psychology of Religion and Synchronicity*, 54.
16. Jung, "Synchronicity: An Acausal Connecting Principle," in *Collected Works* 8:436
17. Jung, "On the Nature of the Psyche," 216.
18. Jung, "On the Nature of the Psyche," 215.
19. Jung, "On the Nature of the Psyche," 215.
20. Jung, "On the Nature of the Psyche," 233.
21. Jung, "On the Nature of the Psyche," 233.
22. Jung, "On the Nature of the Psyche," 207 [emphasis added].
23. Sheldrake, "Is the Sun Conscious?", 23.
24. Sheldrake, "Is the Sun Conscious?", 21.
25. Aziz, *C. G. Jung's Psychology of Religion and Synchronicity*, 54.
26. Jung, "Synchronicity: An Acausal Connecting Principle," 436.
27. Jung, "On the Nature of the Psyche," 213 [emphasis added].
28. Jung, "On the Nature of the Psyche," 230 [emphasis added].
29. Jung, *Letters*, 2:317–19.
30. Jung, "On the Nature of the Psyche," 207–8; Jung, *Collected Works* 6:797 [emphasis added].
31. Jung, "Synchronicity: An Acausal Connecting Principle," 436 [emphasis added].
32. Jung, "On the Nature of the Psyche," 218.
33. Jung, "On the Nature of the Psyche," 207–8.
34. Jung, "On the Nature of the Psyche," 219.
35. Jung, "On the Nature of the Psyche," 221.
36. Jung, "On the Nature of the Psyche," 225.
37. Jung, "On the Nature of the Psyche," 222.
38. Jung, "On the Nature of the Psyche," 223–24.
39. Jung, "On the Nature of the Psyche," 222–23.

CHAPTER 6. TEILHARD'S SENSE OF COLLECTIVE CONSCIOUSNESS

1. Teilhard, *The Human Phenomenon*, 2.

2. Teilhard, *Lettres Intimes*, 269.
3. Teilhard, *The Phenomenon of Man*, xix.
4. Teilhard, *L'oeuvre Scientifique*.
5. Cheney, "Has Teilhard de Chardin 'Really' Joined the Within and the Without of Things?," 217.
6. Cheney, "Has Teilhard de Chardin 'Really' Joined the Within and the Without of Things?," 217.
7. Teilhard, "The Stuff of the Universe," 383. Essay written in 1953.
8. King, *Spirit of Fire: The Life and Vision of Teilhard de Chardin*, 1.
9. King, *Spirit of Fire*, 4.
10. King, *Spirit of Fire*, 4.
11. Aczel, *The Jesuit and the Skull*, 72.
12. Aczel, *The Jesuit and the Skull*, 24.
13. Teilhard, *The Heart of Matter*, 25.
14. Teilhard, *The Heart of Matter*, 74.
15. Raven, *Teilhard de Chardin: Scientist and Seer*, 164–65.
16. Bergson, *Creative Evolution*.
17. Teilhard, *The Human Phenomenon*, 149.
18. Teilhard, *The Heart of Matter*, 25.
19. Teilhard, *Letters to Leontine Zanta*, 102.
20. King, *Spirit of Fire*, 38.
21. King, *Spirit of Fire*, 47.
22. King, *Spirit of Fire*, 49.
23. King, *Pierre Teilhard de Chardin*, 52.
24. Aczel, *The Jesuit and the Skull*, 82.
25. Aczel, *The Jesuit and the Skull*, 81.
26. Aczel, *The Jesuit and the Skull*, 77.
27. Corte, *Pierre Teilhard de Chardin: His Life and Spirit*, 15.
28. Teilhard, "The Making of a Mind: Letters from a Soldier-Priest, 1914–1919," 205.
29. Teilhard, "Nostalgia for the Front," 172.
30. Teilhard, "Nostalgia for the Front," 172.
31. Horne, *The Price of Glory: Verdun 1916*, 328.
32. Teilhard, "Christ in Matter," 61. Essay written in 1916.
33. Tudzynski, Correia, and Keller, "Biotechnology and Genetics of Ergot Alkaloids."
34. Teilhard, "Christ in Matter," 61–65. Essay written 1916.
35. Teilhard, "The Christic," in *The Heart of Matter*, 83. Essay written in 1955.
36. Teilhard, "The Christic," 83.
37. Teilhard, "The Christic," 82–83 [emphasis added].
38. King, *Spirit of Fire*, 59.

39. King, *Spirit of Fire*, 116.
40. Teilhard, "Human Energy," 118. Essay written in 1937.
41. Teilhard, "My Universe," 197. Essay written April 14, 1918.
42. Teilhard, *The Divine Milieu*, 76–77.
43. Aczel, *The Jesuit and the Skull*, 123–24.
44. Aczel, *The Jesuit and the Skull*, 124.
45. Aczel, *The Jesuit and the Skull*, 132.
46. Aczel, *The Jesuit and the Skull*, 132.
47. Association for Asian Studies, *Annals*, 51.
48. King, *Spirit of Fire*, 233–34.
49. Teilhard, *The Human Phenomenon*, 2.
50. Teilhard, "Life and the Planets," 123. Lecture delivered in 1945.
51. Teilhard, "Life and the Planets," 122.
52. Teilhard, "My Fundamental Vision," 164. Essay written in 1948.
53. Teilhard, "My Fundamental Vision," 83. Note that the Eocene epoch lasted from 56 to 33.9 million years ago.
54. Teilhard, "My Fundamental Vision," 84.
55. Morgan, *Emergent Evolution: Gifford Lectures, 1921–22*.
56. Haisch, *The Purpose-Guided Universe: Believing in Einstein, Darwin, and God*.
57. King, *Spirit of Fire*, 233–34.
58. King, *Spirit of Fire*, 233–34.
59. Leroy, "Teilhard de Chardin: The Man," 32.
60. Teilhard, *Letters from a Traveler*, 291
61. Teilhard, "The Zest for Living," 231.
62. King, *Pierre Teilhard de Chardin*, 17.
63. Teilhard, "Centrology: An Essay in the Dialectic of Union." Written in 1944.
64. King, *Spirit of Fire*, 213.
65. Teilhard, "The Great Monad," 182. Essay written in 1918.
66. Teilhard, *The Heart of Matter*.
67. King, *Spirit of Fire*, 84.
68. Speaight, *The Life of Teilhard de Chardin*, 117.
69. Teilhard, as quoted in Cuénot, *Teilhard de Chardin: A Biographical Study*, 59.
70. King, *Spirit of Fire*, 84.
71. Bailes, *Science and Russian Culture in an Age of Revolutions*.
72. Samson and Pitt, *The Biosphere and Noosphere Reader*, 94–95.
73. Vernadsky, *The Biosphere*, 16.
74. Aczel, *The Jesuit and the Skull*, 86.
75. Teilhard, "Hominization," 61.
76. Teilhard, "Hominization," 62.

77. Teilhard, "Hominization," 73–78.
78. Teilhard, "The Death-Barrier and Co-Reflection," 402.
79. Teilhard, "The Death-Barrier and Co-Reflection," 78.
80. Teilhard, *The Divine Milieu*, 25.
81. Teilhard, *The Divine Milieu*, 24.
82. Teilhard, *The Divine Milieu*, 26.
83. Teilhard, *The Divine Milieu*, 128–129.
84. Teilhard, *The Divine Milieu*, 120.
85. Teilhard, *Christianity and Evolution*, 160
86. Teilhard, *The Divine Milieu*, 128.
87. Teilhard, *The Divine Milieu*, 180.
88. Teilhard, "The Christic."
89. Teilhard, "The Christic," 82.
90. Teilhard, "The Christic," 90.
91. King, *Spirit of Fire*, 97.
92. King, *Spirit of Fire*, 98.
93. King, *Spirit of Fire*, 106.
94. King, *Spirit of Fire*, 106–8.
95. Teilhard, "Letters to Two Friends 1926–1952," 5.
96. Aczel, *The Jesuit and the Skull*, 78.
97. Aczel, *The Jesuit and the Skull*, 78.
98. Aczel, *The Jesuit and the Skull*, 79.
99. King, *Spirit of Fire*, 93.
100. Cuénot, *Teilhard de Chardin: A Biographical Study*, 257.
101. Cuénot, *Teilhard de Chardin: A Biographical Study*, 258.
102. Leckie, *Delivered from Evil: The Saga of World War II*.
103. Aczel, *The Jesuit and the Skull*, 213.
104. King, *Spirit of Fire*, 230.
105. Aczel, *The Jesuit and the Skull*, 221.
106. King, *Spirit of Fire*, 230.
107. Aczel, *The Jesuit and the Skull*.
108. Aczel, *The Jesuit and the Skull*, 231.
109. Dunwell, *The Hudson: America's River*, 140.

CHAPTER 7. HYPERPHYSICS:
SPECULATIVE PHYSICS BEYOND MATERIAL SCIENCE

1. Teilhard de Chardin, *Activation of Energy*, 383.
2. Teilhard, "Centrology: An Essay in a Dialectic of Union," 99.

3. McFadden, "Synchronous Firing and Its Influence on the Brain's Electromagnetic Field: Evidence for an Electromagnetic Field Theory of Consciousness," 23.
4. Pockett, *The Nature of Consciousness: A Hypothesis*, 7.
5. Teilhard, *Human Phenomenon*, 2.
6. Radhakrishnan, *History of Philosophy Eastern and Western*, 57.
7. Skrbina, *Panpsychism in the West*, 30.
8. MacKenna, *Plotinus: The Enneads*.
9. Radhakrishnan, *History of Philosophy Eastern and Western*, 115.
10. MacKenna, *Plotinus: The Enneads*, 712.
11. Jung, *Psychology and Alchemy*.
12. Teilhard, "Centrology," 127.
13. Teilhard, as quoted in Cuénot, *Teilhard de Chardin: A Biographical Study*, 59.
14. Teilhard, "The Convergence of the Universe," 285. Essay written in 1951.
15. Samson and Pitt, *The Biosphere and Noosphere Reader*, xi.
16. Samson and Pitt, *The Biosphere and Noosphere Reader*, 2–3.
17. Teilhard, "The Spirit of the Earth," 42. Essay written in 1931.
18. Teilhard, "The Phenomenon of Spirituality," 96–97.
19. Teilhard, "The Nature of the Point Omega," 160.
20. Teilhard, "From Cosmos to Cosmogenesis," 257. Essay written in 1951.
21. Teilhard, "Centrology," 103–106.
22. Teilhard, "The Phenomenon of Spirituality," 93.
23. Teilhard, "The Phenomenon of Spirituality," 93.
24. Teilhard, "The Phenomenon of Spirituality," 93.
25. Teilhard, "The Phenomenon of Spirituality," 93–94.
26. Teilhard, "The Phenomenon of Spirituality," 93–94.
27. Teilhard, "Human Energy," 130–31.
28. Teilhard, "The Phenomenon of Spirituality," 99.
29. Teilhard, "The Phenomenon of Spirituality," 98.
30. Teilhard, "The Phenomenon of Spirituality," 98.
31. Teilhard, "The Phenomenon of Spirituality," 101.
32. Teilhard, "The Phenomenon of Spirituality," 103.
33. Teilhard, "The Phenomenon of Spirituality," 104.
34. Teilhard, "The Phenomenon of Spirituality," 106.
35. Teilhard, "The Phenomenon of Spirituality," 105.
36. Teilhard, "The Phenomenon of Spirituality," 106.
37. Teilhard, "The Phenomenon of Spirituality," 106.
38. Teilhard, "The Phenomenon of Spirituality," 105.
39. Teilhard, "The Phenomenon of Spirituality," 107.
40. Teilhard, "The Phenomenon of Spirituality," 107.

41. Teilhard, "The Phenomenon of Spirituality," 107–8.
42. Teilhard, "The Phenomenon of Spirituality," 108.
43. Teilhard, "The Phenomenon of Spirituality," 109–11.
44. Teilhard, "The Phenomenon of Spirituality," 109–11.
45. Teilhard, "The Phenomenon of Spirituality," 109–11.
46. Teilhard, "The Phenomenon of Spirituality," 99.
47. Teilhard, "The Phenomenon of Spirituality," 101.
48. Teilhard, "The Phenomenon of Spirituality," 103.
49. Joye, "The Pribram-Bohm Holoflux Theory of Consciousness".
50. Joye, "The Pribram-Bohm Holoflux Theory of Consciousness", 93–94.
51. Gao, *Dark Energy: From Einstein's Biggest Blunder to the Holographic Universe*.
52. Teilhard, "The Christic," in *The Heart of Matter*, 82.
53. Teilhard. "The Divine Milieu" in *The Heart of Matter*, 49–50.
54. Teilhard, *The Heart of Matter*, 50.
55. Teilhard, *The Heart of Matter*, 49.
56. Teilhard, *Christianity and Evolution*, 173–74.
57. Teilhard, *Christianity and Evolution*, 184.
58. Teilhard, *Christianity and Evolution*, 185.
59. Teilhard, *Christianity and Evolution*, 186.

CHAPTER 8. THE ENERGETIC NOOSPHERE

1. Teilhard, "The Energy of Evolution," 361–62. Essay written in 1953.
2. Teilhard, "The Activation of Human Energy," 393.
3. Teilhard, *Activation of Energy*.
4. Teilhard, "The Atomism of Spirit," 29. Essay written in 1941.
5. Teilhard, *The Human Phenomenon*, 109.
6. Teilhard, "The Zest for Living," 242.
7. Teilhard, "The Zest for Living," 242.
8. Teilhard, "The Activation of Human Energy," 393.
9. Teilhard, "Centrology," 121n10.
10. Teilhard, "Centrology," 120.
11. Teilhard, *Christianity and Evolution*, 56; and *The Appearance of Man*, 33.
12. Samson and Pitt, *The Biosphere and Noosphere Reader*, 3.
13. Allaby and Allaby, *A Dictionary of Earth Sciences*, 72.
14. Malinski, *Chemistry of the Heart*, 61.
15. The Bolshakovo transmitter near Bolshakovo, Russia is credited as the most powerful medium-wave broadcasting station. This station gave a 2,500,000 Watt worldwide voice to the Voice of Russia on medium-wave (1116 kHz and 1386 kHz).

16. Walker, *Three Mile Island: A Nuclear Crisis in Historical Perspective*, 12.
17. McCraty, Deyhle, and Childre, "The Global Coherence Initiative: Creating a Coherent Planetary Standing Wave," 75.
18. McCraty, Deyhle, and Childre, "The Global Coherence Initiative: Creating a Coherent Planetary Standing Wave," 76.
19. Teilhard, *Christianity and Evolution: Reflections on Science and Religion*, 231.
20. Teilhard, "Human Energy."
21. Teilhard, "Human Energy," 117.
22. Teilhard, "Human Energy," 118; emphasis added.
23. Teilhard, "Human Energy," 128.
24. Teilhard, "Human Energy," 129–30.
25. Jung, "On the Nature of the Psyche," in *The Structure and Dynamics of the Psyche*, 207.
26. Teilhard, "The Spirit of the Earth," 35.
27. Teilhard, "The Spirit of the Earth," 33.
28. Teilhard, "Human Energy," 130–31.
29. Teilhard, "Human Energy," 130–31.
30. Teilhard, "Human Energy," 130–31.
31. Teilhard, "Human Energy," 138.
32. Teilhard, "Human Energy," 141.
33. Teilhard, "Cosmic Life," 15.
34. Teilhard, "Human Energy," 143.
35. Teilhard, "Human Energy," 143–44.
36. Teilhard, "Human Energy," 144.
37. Teilhard, "Human Energy," 145.
38. de Lubac, *The Religion of Teilhard de Chardin*, 123.
39. Teilhard, "Life and the Planets."
40. Teilhard, "Life and the Planets," 122.
41. Teilhard, *The Human Phenomenon*, 191–93.
42. Teilhard, "Human Energy," 138.
43. Teilhard, "The Activation of Human Energy," 393.
44. Teilhard, "The Nature of the Point Omega," 272–73.
45. Teilhard, "Human Energy," 162.
46. Teilhard, "Centrology," 99.
47. Teilhard, "Centrology," 100.
48. Teilhard, "Centrology," 101.
49. Teilhard, "Centrology," 102.
50. Teilhard, "Centrology," 102n1.
51. Teilhard, "Centrology," 102.

52. Teilhard, "Centrology," 103.
53. Teilhard, "Man's Place in the Universe," 226.
54. Teilhard, "The Atomism of Spirit," 40.
55. Teilhard, *The Human Phenomenon*, 110.
56. Teilhard, "Centrology," 103.
57. Teilhard, "Universalization and Union," 91.
58. Teilhard, "Centrology," 120.
59. Teilhard, "Centrology," 110.
60. Teilhard, "Centrology," 110.
61. Teilhard, "Outline of a Dialectic of Spirit," 144.
62. Teilhard, "Centrology," 103.
63. Rescher, *G. W. Leibniz's Monadology*.
64. Teilhard, "Centrology," 104.
65. Teilhard, "Centrology," 105.
66. Teilhard, "Centrology," 106.
67. Teilhard, "Centrology," 107.
68. Teilhard, "Centrology," 108.
69. Teilhard, "Centrology," 109.
70. Sheldrake, *A New Science of Life: The Hypothesis of Morphic Resonance*.
71. Teilhard, "Centrology," 109.
72. Teilhard, "Centrology," 109.
73. Teilhard, "Centrology," 109.
74. Teilhard, "Centrology," 110.
75. Teilhard, "Centrology," 110.
76. Teilhard, "Centrology," 110–11.
77. Teilhard, "Centrology," 111.
78. Teilhard, "Centrology," 111.
79. Teilhard, "The Phenomenon of Spirituality," 100.
80. Teilhard, "The Phenomenon of Spirituality," 99.
81. Teilhard, *The Phenomenon of Man*, 73–74.
82. Teilhard, "Centrology," 112.
83. Teilhard, "Centrology," 112.
84. Teilhard, "Centrology," 112.
85. Teilhard, "Centrology," 113.
86. Teilhard, "Centrology," 113.
87. Teilhard, "Centrology," 113.
88. Assistance with these Latin translations was provided by Fr. Thomas Matus, Ph.D., a Camaldolese Benedictine monk, in an email message to author, August 27, 2015.
89. Teilhard, "Centrology," 114.

90. Teilhard, "Centrology," 114.
91. Teilhard, "Centrology," 114.
92. Teilhard, "Centrology," 114.
93. Teilhard, "The Formation of the Noosphere II," 111. Essay written in 1949.
94. Persinger, Booth, and Koren, "Increased Feelings of the Sensed Presence."
95. Teilhard, "Centrology," 114–15.
96. Teilhard, "Centrology," 116.
97. Teilhard, "Centrology," 115.
98. Joye, "The Pribram-Bohm Holoflux Theory of Consciousness".
99. Joye, "The Pribram-Bohm Holoflux Theory of Consciousness". 119.
100. Joye, "The Pribram-Bohm Holoflux Theory of Consciousness". 116.
101. Teilhard, "Hominization," in *The Vision of the Past*, 78.
102. Teilhard, "Centrology," 122.
103. Teilhard, "Centrology," 116–17.
104. Teilhard, "Centrology," 100.
105. Teilhard, "Centrology," 102.
106. Goswami, *The Visionary Window*.
107. Gebser, *The Ever-Present Origin*, 37.
108. Gebser, *The Ever-Present Origin*, 39.
109. Teilhard, "Centrology," 117.
110. Teilhard, "Centrology," 117.
111. Einstein, *Autobiographical Notes*, 17.
112. Teilhard, "The Death-Barrier and Co-Reflection," 403.
113. de Terra, *Memories of Teilhard de Chardin*, 42.
114. Teilhard, "The Death-Barrier and Co-Reflection," 402.

CHAPTER 9. THE SUPERPOSITION OF CONSCIOUSNESS

1. Kuo, *Network Analysis and Synthesis*, 13.
2. Wiener, *Cybernetics: Or Control and Communication in the Animal and the Machine*, 202.
3. Bohm, *Quantum Theory*, 1.
4. James, *A Pluralistic Universe*, 318.

CHAPTER 10. THE PRIBRAM-BOHM HOLOFLUX THEORY

1. Pribram, *The Form Within*, 524.
2. Pribram, "Brain and Mathematics," 13.
3. Pribram, "Consciousness Reassessed."
4. Bohm and Hiley, *The Undivided Universe*, 382.

5. Pribram, "Brain and Mathematics," 219.
6. Pribram, *Brain and Perception: Holonomy and Structure*, 19.
7. Pribram, *Brain and Perception*, 142.
8. Pribram, "What the Fuss Is All About," 29.
9. Pribram, "Prolegomenon for a Holonomic Brain Theory."
10. Pribram, *Brain and Perception*, 73.
11. Pribram, *The Form Within*, 535–36.
12. *The Stanford Daily*, Monday, September 26, 1977, 8.
13. Dawson, *Cults and New Religious Movements*, 54.
14. Pribram, *Brain and Perception*, 272–73.
15. Peat, *Infinite Potential*, 193–94.
16. Peat, *Infinite Potential*, 193–94.
17. Bohm, *Wholeness and the Implicate Order*, 188.
18. Bohm, *Wholeness and the Implicate Order*, 189.
19. Bohm, *Wholeness and the Implicate Order*, 189.
20. Bohm, *Wholeness and the Implicate Order*, 172.
21. Carr, *Universe or Multiverse?*, 13.
22. Carr, *Universe or Multiverse?*, 13.
23. Read, *From Alchemy to Chemistry*, 179–80.
24. Bohm and Hiley, *The Undivided Universe*, 381–82.
25. Bohm and Weber, "Nature as Creativity," 35–36.
26. Pribram, "Brain and Mathematics," 13.
27. Pribram, *Brain and Perception*, 70.
28. Spencer-Brown, *Laws of Form*, 105.
29. Bohm, *Wholeness and the Implicate Order*, 190.
30. Bohm, *Wholeness and the Implicate Order*, 192.
31. Bohm, *Wholeness and the Implicate Order*, 196–97.
32. Bohm, *Wholeness and the Implicate Order*, 190-91.

CHAPTER 11. THE DIGITAL UNIVERSE

1. Smolin, *Time Reborn*.
2. Wheeler, *A Journey into Gravity and Spacetime*, 222.
3. Dewey, *The Theory of Laminated Spacetime*.
4. Dewey, *The Theory of Laminated Spacetime*, 95.
5. Bohm, *Wholeness and the Implicate Order*, 35–36.
6. Bohm, *Wholeness and the Implicate Order*, 70.
7. Bohm, *Wholeness and the Implicate Order*, 77.
8. Susskind, *The Black Hole War*.
9. Wheeler, *A Journey into Gravity and Spacetime*, 222.

10. Bekenstein, "Black Holes and Entropy."
11. Romanes, *Cunningham's Textbook of Anatomy*, 137.
12. Feynman, Leighton, and Sands, *The Feynman Lectures on Physics*, 78.
13. Dorf, *Electrical Engineering Handbook*.
14. Dorf, *Electrical Engineering Handbook*.
15. Gowar, 1993, 64.
16. Kuo, *Network Analysis and Synthesis*, 40.
17. Wiener, *Cybernetics*, 198.
18. Wiener, *Cybernetics*, 202.
19. Wiener, *Cybernetics*, 202; emphasis added.
20. Jung, "On the Nature of the Psyche," in *The Structure and Dynamics of the Psyche*, 215.
21. Kuo, *Network Analysis and Synthesis*, 13.

CHAPTER 12. EXPLORING THE HIDDEN DIMENSIONS OF THE MIND

1. McGilchrist, *The Matter with Things*.
2. Krishnamurti and Bohm, "On Intelligence."
3. Grof, "Revision and Re-Enchantument of Psychology," 137.
4. Bohm, *Wholeness and the Implicate Order*, 291.
5. Fechner, *Religion of a Scientist*, 158.

Bibliography

Aczel, Amir D. *The Jesuit and the Skull: Teilhard de Chardin, Evolution, and the Search for Peking Man*. New York: Riverhead Books, 2007.
Addison, Ann. *Jung's Psychoid Concept Contextualised*. Routledge. 2018.
Adelman, George, ed. *Encyclopedia of Neuroscience*. Vol. 1. Boston, MA: Birkhäuser Boston, 1987.
Allaby, Michael, and Ailisa Allaby, eds. *A Dictionary of Earth Sciences*. 2nd ed. New York: Oxford University Press, 1999.
Association for Asian Studies, Southeast Conference, *Annals,* vols. 1–5.
Aziz, Robert. *C. G. Jung's Psychology of Religion and Synchronicity*. Albany: State University of New York Press, 1990.
Bailes, Kendall E. *Science and Russian Culture in an Age of Revolutions: V. I. Vernadsky and His Scientific School*. Indiana: Indiana University Press, 1990.
Bekenstein, Jacob D. "Black Holes and Entropy." *Physical Review* 7, no. 8 (1973): 2333–46.
Bergson, Henri. *Creative Evolution*. Translated by Arthur Mitchell. New York: Henry Holt, 1911.
Blofeld, John. *Tantric Mysticism of Tibet: A Practical Guide to the Theory, Purpose, and Techniques of Tantric Meditation*. Boston, MA: E. P. Dutton, 1970.
———. *The Wheel of Life*. Boulder, CO: Shambhala, 1978.
Bohm, David. "The Enfolding-Unfolding Universe: A Conversation with David Bohm." In *The Holographic Paradigm and Other Paradoxes: Exploring the Leading Edge of Science*, edited by Ken Wilber, 44–104. Boulder, CO: Shambhala, 1978.
———. *Quantum Theory*. New York: Prentice-Hall, 1951.
———. *Unfolding Meaning: A Week of Dialog with David Bohm*. New York: Routledge, 1985.
———. *Wholeness and the Implicate Order*. London: Routledge, 1980.
Bohm, David, and Basil J. Hiley. *The Undivided Universe: An Ontological Interpretation of Quantum Theory*. London: Routledge, 1993.

Bohm, David, and F. David Peat. *Science, Order, and Creativity*. London: Routledge, 1987.
Bohm, David, and Renée Weber. "Nature as Creativity." *ReVision* 5, no. 2 (1982): 35–40.
———. "The Physicist and the Mystic—Is a Dialogue between Them Possible?" In *The Holographic Paradigm and Other Paradoxes*, edited by Ken Wilber, 187–214. Boulder, CO: Shambhala, 1982.
Bucke, Richard Maurice. *Cosmic Consciousness: A Study in the Evolution of the Human Mind*. New York: E. P. Dutton, 1901.
Carr, Bernard. *Universe or Multiverse?* Cambridge: Cambridge University Press, 2007.
Chen, Frances F. *Introduction to Plasma Physics and Controlled Fusion*. Vol. 1, *Plasma Physics*. 2nd ed. New York: Springer, 2006.
Cheney, B. "Has Teilhard de Chardin 'Really' Joined the Within and the Without of Things?" *Sewanee Review* 73, no.2: 217–36.
Churton, Tobias. *Aleister Crowley: The Biography; Spiritual Revolutionary, Romantic Explorer, Occult Master*. London: Watkins Publishing, 2012.
Corte, Nicholas. *Pierre Teilhard de Chardin: His Life and Spirit*. London: Barrie and Rockliffe, 1957.
Crabtree, Adam. *From Mesmer to Freud: Magnetic Sleep and the Roots of Psychological Healing*. New Haven, CT: Yale University Press, 1993.
Cuénot, Claude. *Teilhard de Chardin: A Biographical Study*. London: Burnes & Oates, 1965.
Dawson, Lorne L. *Cults and New Religious Movements*. Malden, MA: Blackwell Publishing, 2003.
de Lubac, Henri. *The Religion of Teilhard de Chardin*. Translated by René Hague. New York: Desclée, 1967.
de Terra, Helmut. *Memories of Teilhard de Chardin*. New York: Harper & Row, 1964.
Dewey, Barbara. *The Theory of Laminated Spacetime*. New York: Bartholomew Books, 1985.
Dorf, Richard C. *The Electrical Engineering Handbook*, 2nd ed. Boca Raton, FL: CRC Press, 1997.
Duffy, Kathleen. *Teilhard's Struggles: Embracing the Work of Evolution*. Maryknoll, NY: Orbis Books, 2019.
Dunwell, Frances. *The Hudson: America's River*. New York: Columbia University Press, 2008.
Dyczkowski, Mark S. G. *A Journey in the World of the Tantras*. India: Indica Books, 2004.
Einstein, Albert. *Autobiographical Notes*. Peru, IL: Carus Publishing, 1979.
Eliade, Mircea. *The Road to the Center*. Bucharest: University of Bucharest, 1991.
———. *Yoga: Immortality and Freedom*. Princeton, NJ: Princeton University Press, 1954.
Fechner, Gustav Theodor. *Über die Dinge des Himmels und des Jenseits* (Concerning Matters of Heaven and the World to Come). 3 vols. Leipzig, 1851.
———. *Elemente der Psychophysik*. Leipzig: Breitkopf und Härtel, 1860.
———. *The Little Book of Life after Death*. Boston, MA: Little, Brown, 1905.
———. *Nanna, or the Soul-Life of Plants*. Leipzig: Leopold Voss, 1921.

———. *Religion of a Scientist*. New York: Pantheon Books, 1946.
Feynman, Richard, Robert Leighton, and Matthew Sands. *The Feynman Lectures on Physics*. Vol. 1. Boston, MA: Addison-Wesley, 1964.
Fox, Charles R. "Gustav Fechner: The Man Who Introduced Soul to Science." Academia Letters website, Article 38. 2020.
Gao, Shan. *Dark Energy: From Einstein's Biggest Blunder to the Holographic Universe*. Seattle: Amazon Kindle Direct, 2014.
Gebser, Jean. *The Ever-Present Origin: Part One: Foundations of the Aperspectival World*. Translated by J. Keckeis. Stuttgart, Germany: Deutsche Verlags-Anstalt, 1949.
Goswami, Amit. *The Visionary Window: A Quantum Physicist's Guide to Enlightenment*. Wheaton, IL: Quest, 2000.
Gowar, John. *Optical Communication Systems*. New York: Prentice-Hall, 1984.
Grof, Stanislav. "Revision and Re-Enchantument of Psychology: Legacy of Half a Century of Consciousness Research." *Journal of Transpersonal Psychology* 44, no. 2 (2012):137–63.
Gurdjieff, G. I. *Beelzebub's Tales to His Grandson, or An Objectively Impartial Criticism of the Life of Man*. New York: Harcourt, 1950.
Haisch, Bernard. *The Purpose-Guided Universe: Believing in Einstein, Darwin, and God*. Franklin Lakes, NJ: Career, 2010.
HeartMath Institute. "Energetic Communication." Chapter 6 of *Science of the Heart: Exploring the Role of the Heart in Human Performance*. HeartMath Institute website. Accessed September 19, 2024.
Heidelberger, Michael. *Nature from Within: Gustav Theodor Fechner and His Psychophysical Worldview*. Pittsburgh, PA: University of Pittsburgh Press, 2004.
Hoffman, Donald. "Donald Hoffman-Consciousness, Mysteries Beyond Spacetime, and Waking up from the Dream Life,"Posted to Youtube May 30, 2024, by The Weekend University.
Horne, Alistair. *The Price of Glory: Verdun 1916*. New York: St. Martin's, 1962.
Huxley, Aldous. *The Doors of Perception*. New York: Harper & Brothers, 1954.
Irenaeus. *Against Heresies*. Ashland, OR: Beloved Publishing, 2014.
James, William. *A Pluralistic Universe*. New York: Longmans, Greene, 1909.
———. *The Principles of Psychology*. Mineola, NY: Dover, 1950. First published 1890.
———. *The Varieties of Religious Experience*. New York: Longmans, Greene, 1902.
Jones, Roger S. *Physics for the Rest of Us: Ten Basic Ideas of Twentieth-Century Physics That Everyone Should Know . . . and How They Have Shaped Our Culture and Consciousness*. New York: Barnes and Noble, 1999.
Joye, Shelli R. *The Electromagnetic Brain: EM Theories on the Nature of Consciousness*. Rochester, VT: Inner Traditions, 2020.
———. *The Little Book of Consciousness: Holonomic Brain Theory and the Implicate Order*. Viola, CA: Viola Institute, 2017.

———. "The Pribram-Bohm Holoflux Theory of Consciousness: An Integral Interpretation of the Theories of Karl Pribram, David Bohm, and Pierre Teilhard de Chardin." Ph.D. diss., California Institute of Integral Studies, 2016. ProQuest (10117892).

———. "The Pribram-Bohm Hypothesis." *Consciousness: Ideas and Research for the Twenty-First Century* 3, no. 3 (2016): article 1.

———. *Sub-Quantum Consciousness: A Geometry of Consciousness Based upon the Work of Karl Pribram, David Bohm, and Pierre Teilhard De Chardin.* Viola, CA: Viola Institute, 2019.

———. *Teilhard's Hyperphysics: Energy and the Noosphere.* Viola, CA: Viola Institute, 2020.

———. *Tuning the Mind: The Geometries of Consciousness.* Viola, CA: Viola Institute, 2017.

Jung, C.G. *The Collected Works of C. G. Jung*, Vol. 13. Princeton University Press, 1953.

———. *C.G. Jung Letters*, Vol. 2. Princeton University Press, 2021.

———. "The Structure and Dynamics of the Psyche." In *The Collected Works of C.G. Jung*, vol. 8. Translated by R. F. C. Hull. Princeton, NJ: Princeton University Press, 1960.

———. "Synchronicity: An Acausal Connecting Principle," vol. 8 of *Collected Works*, 436.

Jung, C. G., and M.-L. von Franz, eds. *Man and His Symbols.* New York: Random House, 1964.

Jung, C. G., and W. Pauli. *The Interpretation of Nature and the Psyche. C.G. Jung-Synchronicity: An Acausal Connecting Principle.* Pantheon, 1955.

Jung, C. G., Sonu Shamdasani, Ulrich Hoerni, Mark Kyburz, and John Peck. *The Red Book = Liber Novus: A Reader's Edition.* New York: W. W. Norton, 2012.

Kahneman, Daniel. *Thinking, Fast and Slow.* New York: Farrar, Straus, and Giroux, 2011.

King, Ursula. *Pierre Teilhard de Chardin: Writings Selected with an Introduction by Ursula King.* New York: Orbis Books, 1999.

———. *Spirit of Fire: The Life and Vision of Teilhard de Chardin.* New York: Orbis Books, 1996.

Krishnamurti, J., and David Bohm. "On Intelligence." In J. Krishnamurti, *The Awakening of Intelligence*, 477–507. New York: Harper & Row, 1973.

Kuo, F. *Network Analysis and Synthesis.* Murray Hill, NJ: Bell Telephone Labs, 1962.

Leary, Timothy, Ralph Metzner, and Richard Alpert. *The Psychedelic Experience: A Manual Based upon the Tibetan Book of the Dead.* New York: University Books, 1964.

Leckie, Robert. *Delivered from Evil: The Saga of World War II.* New York: Harper Perennial, 1988.

Leroy, Pierre. "Teilhard de Chardin: The Man." Introduction to *The Divine Milieu*, by Teilhard de Chardin, 13–42. New York: Harper & Row, 1960.

Lilly, John. *The Scientist: A Metaphysical Autobiography.* Berkeley, CA: Ronin, 1988.

Lindorff, D. *Pauli and Jung: The Meeting of Two Great* Minds. Wheaton, IL: Quest Books, 2004.

MacKenna, Stephen. *Plotinus: The Enneads*. Burdett, New York: Larson Publications, 1992.

———. *Psychology and Alchemy*. Vol. 8. Translated by R. F. C. Hull. Princeton, NJ: Princeton University Press, 1953.

Malinski, Tadeusz. *Chemistry of the Heart*. Ohio: Biochemistry Research Laboratory, 1960. Retrieved from hypertextbook website.

May, Brian, and Garik Israelian. *STARMUS: 50 Years of Man in Space*. Prague: Starmus, 2014.

McCraty, Rollin, Annette Deyhle, and Doc Childre. 2012. "The Global Coherence Initiative: Creating a Coherent Planetary Standing Wave." *Global Advances in Health and Medicine* 1, no.1: 64–77. Retrieved from Heartmath website.

McFadden, Johnjoe. "Synchronous Firing and Its Influence on the Brain's Electromagnetic Field: Evidence for an Electromagnetic Field Theory of Consciousness." *Journal of Consciousness Studies* 9, no. 4 (2002): 23–50.

McGilchrist, Iain. *The Matter with Things: Our Brains, Our Delusions, and the Unmaking of the World*. London: Perspectiva Press, 2021.

Minkowski, Hermann. "The Fundamental Equations for Electromagnetic Processes in Moving Bodies." In *The Principle of Relativity*. 1–69. Calcutta: University Press, 1920.

Mishra, Ramamurti. *Yoga Sutras: The Textbook of Yoga Psychology*. New York: Julian Press, 1963.

Morgan, Conway Lloyd. *Emergent Evolution: Gifford Lectures, 1921–22*. New York: Simon & Schuster, 1978.

Ouspensky, P. D. *In Search of the Miraculous: The Definitive Exploration of G. I. Gurdjieff's Mystical Thought and Universal View*. New York: Harcourt, Brace, 1949.

Peat, F. David. *Infinite Potential: The Life and Times of David Bohm*. Reading, MA: Addison-Wesley, 1997.

Persinger, Michael A., J. C. Booth, and S. A. Koren. "Increased Feelings of the Sensed Presence and Increased Geomagnetic Activity at the Time of the Experience during Exposures to Transcerebral Weak Complex Magnetic Fields." *International Journal of Neuroscience* 115 (2005): 1039–65.

Pockett, Susan. *The Nature of Consciousness: A Hypothesis*. Lincoln, NE: Writers Club, 2000.

Pribram, Karl. "What the Fuss Is All About." In *ReVision: A Journal of Consciousness and Transformation*. San Francisco: ReVision Publishing, Spring 1978.

Pribram, Karl H. "Brain and Mathematics." In *Brain and Being: At the Boundary Between Science, Philosophy, Language and Arts*, edited by Gordon Globus, Karl Pribram, and Giuseppe Vitiello, 215–40. Philadelphia, PA: John Benjamins, 2004.

———. *Brain and Perception: Holonomy and Structure in Figural Processing*. New Jersey: Lawrence Erlbaum, 1991.

———. "Consciousness Reassessed." *Mind and Matter* 2, no. 1 (2004): 7–35.

———. *The Form Within: My Point of View*. Westport, CT: Prospecta, 2013.

———. *Languages of the Brain: Experimental Paradoxes and Principles in Neuropsychology.* Saddle River, NJ: Prentice-Hall, 1971.

———. "Prolegomenon for a Holonomic Brain Theory." In *Synergetics of Cognition*, edited by H. Haken. Berlin: Springer-Verlag, 1990.

———. "What Is Mind That the Brain May Order It?" *Proceedings of Symposia in Applied Mathematics: Proceedings of the Norbert Wiener Centenary Congress* 52, 301–29. East Lansing: Michigan State University, 1997.

Radhakrishnan, Sarvepalli. *History of Philosophy Eastern and Western.* London: Allen and Ulwin, 1952.

Raven, Charles E. *Teilhard de Chardin: Scientist and Seer.* London: William Collins Sons, 1962.

Read, John. *From Alchemy to Chemistry.* Mineola, NY: Dover Publication, 1957.

Rescher, Nicholas, ed. *G. W. Leibniz's Monadology. An Edition for Students*, Pittsburgh, PA: University of Pittsburgh Press, 1991.

Romanes, G. J., ed. *Cunningham's Textbook of Anatomy.* 10th ed. London: Oxford Press, 1964.

Samson, Paul R., and David Pitt, eds. *The Biosphere and Noosphere Reader: Global Environment, Society and Change.* New York: Routledge, 1999.

Schroll, M. A., "Understanding Bohm's Holoflux: Clearing up a Conceptual Misunderstanding of the Holographic Paradigm and Clarifying Its Significance to Transpersonal Studies of Consciousness." *Journal of Transpersonal Studies* 32, no. 1 (2013): 132, online

Sheldrake, Rupert. "Is the Sun Conscious?" *Journal of Consciousness Studies* 28 (2021).

———. *Morphic Resonance: The Nature of Formative Causation.* Rochester, VT: Park Street Press, 1989.

———. *A New Science of Life: The Hypothesis of Morphic Resonance.* Rochester, VT: Park Street Press, 1981.

Sher, Leo. "Neuroimaging, Auditory Hallucinations, and the Bicameral Mind." *Journal of Psychiatry & Neuroscience* 25, no. 3 (2000): 239–40.

Skrbina, David. *Panpsychism in the West.* Cambridge, MA: MIT Press, 2007.

Smolin, Lee. *Time Reborn.* London: Penguin Books, 2013.

Snow, C. P. *The Two Cultures.* London: Oxford Press, 1959.

Speaight, Robert. *The Life of Teilhard de Chardin.* New York: Harper & Row, 1967.

Spencer-Brown, G. *Laws of Form.* London: Bohmeier Verlag, 2008.

Steiner, Rudolf. *The Archangel Michael: His Mission and Ours.* Hudson, NY: Anthroposophic Press, 1994.

———. *Knowledge of the Higher Worlds and Its* Attainment. 3rd ed. Translated by George Metaxa. Hudson, NY: Anthroposophic Press, 1947.

Stuart, C., Y. Takahashi, and H. Umezawa. "Mixed-System Brain Dynamics: Neural Memory as a Macroscopic Ordered State." *Foundations of Physics* 9, no. 3–4 (1979): 301–27.

Susskind, Leonard. *The Black Hole War: My Battle with Stephen Hawking to Make the World Safe for Quantum Mechanics.* New York: Little, Brown, 2008.

Sutter, Paul. "How the Universe Could Possibly Have More Dimensions." Space website. February 21, 2020.

Taimni, I. K. *The Science of Yoga.* India: Theosophical Publishing House, 1961.

Tarbuck, Edward J., Frederick K. Lutgens, and Dennis Tasa. *Earth Science.* London: Pearson, 2017.

Teilhard de Chardin, Pierre. *Activation of Energy.* Translated by René Hague. London: William Collins Sons, 1976.

———. "The Activation of Human Energy." In *Activation of Energy*, translated by René Hague, 359–93. London: William Collins Sons, 1976. First published 1953.

———. *The Appearance of Man*, translated by J.M. Cohen. New York: Harper & Row, 1956.

———. "The Atomism of Spirit." In *Activation of Energy*, translated by René Hague, 21–57. London: William Collins Sons, 1976. First written 1941.

———. "Centrology: An Essay in a Dialectic of Union." In *Activation of Energy*, translated by René Hague, 97–127. London: William Collins Sons, 1976. First written 1944.

———. *Christianity and Evolution: Reflections on Science and Religion.* Translated by René Hague. London: William Collins Sons, 1971.

———. "Christ in Matter (1919)." In *The Heart of Matter*, translated by René Hague. London: William Collins Sons, 1953.

———. "The Convergence of the Universe." In *Activation of Energy*, translated by René Hague, 281–96. London: William Collins Sons, 1976. First written 1951.

———. "Cosmic Life (1916)." In *Writings in Time of War*, translated by René Hague, 14–71. London: William Collins Sons, 1968.

———. "The Death-Barrier and Co-Reflection, or the Imminent Awakening of Human Consciousness to the Sense of Its Irreversibility." In *Activation of Energy*, translated by René Hague, 395–406. London: William Collins Sons, 1976. First written 1955.

———. *The Divine Milieu.* New York: Harper & Row, 1960.

———. "The Energy of Evolution." In *Activation of Energy*, translated by René Hague, 359–72. London: William Collins Sons, 1976. First published 1953.

———. "The Formation of the Noosphere." In *Man's Place in Nature: The Human Zoological Group*, translated by René Hague, 96–121. New York: Harper & Row, 1956.

———. "From Cosmos to Cosmogenesis (1951)." In *Activation of Energy*, translated by René Hague. London: William Collins Sons, 1971.

———. "The Great Monad." In *The Heart of Matter*, translated by René Hague, 182–95. New York: Harcourt Brace Jovanovich, 1978. First published 1918.

———. *The Heart of Matter.* Translated by René Hague. New York: Harcourt Brace Jovanovich, 1978.

———. "Hominization." In *The Vision of the Past*, translated by J. M. Cohen, 51–79. New York: Harper & Row, 1966. First written 1923.

———. "Human Energy," In *Human Energy*, translated by J. M. Cohen, 93–112. New York: Harcourt Brace Jovanovitch, 1969.

———. *The Human Phenomenon*. Translated and edited by Sarah Appleton-Weber. Portland, OR: Sussex Academic, 2003. First published 1955.

———. *Letters from a Traveler*. New York: Harper & Row, 1962.

———. *Letters to Leontine Zanta*. Translated by Bernard Wall. New York: Harper & Row, 1969.

———. "Letters to Two Friends (1926–1952)." In *Letters of Teilhard de Chardin*, translated by René Hague. London: William Collins Sons, 1953.

———. *Lettres Intimes de Teilhard de Chardin a Auguste Valensin, Bruno de Solages, et Henri de Lubac 1919–1955*. Paris: Aubier Montaigne, 1972.

———. "Life and the Planets." In *The Future of Man*, translated by Norman Denny, 97–123. New York: Harper & Row, 1959. First published 1945.

———. *L'Oeuvre Scientifique*. Edited by Nicole and Karl Schmitz-Moormann. 10 vols. Munich: Walter-Verlag, 1971.

———. *The Making of a Mind: Letters from a Soldier-Priest* (1914–1919). London: William Collins Sons, 1976.

———. "My Fundamental Vision." In *Toward the Future*, translated by René Hague, 163–208. London: William Collins Sons, 1975. First published 1948.

———. "My Universe (1924)." In *The Heart of Matter*, translated by René Hague. London: William Collins Sons, 1953.

———. "Nostalgia for the Front." In *The Heart of Matter*, translated by René Hague, 168–81. New York: Harcourt Brace Jovanovich, 1978. First published 1917.

———. "The Nature of the Point Omega (1954)." In *The Appearance of Man*, translated by J.M. Cohen, 93–112. London: William Collins Sons, 1956.

———. "Outline of a Dialectic of Spirit (1946)." In *Activation of Energy*, translated by René Hague. London: William Collins Sons, 1971.

———. *The Phenomenon of Man*. Translated by Bernard Wall. New York: Harper & Row, 1959.

———. "The Phenomenon of Spirituality (1937)." In *Human Energy*, translated by J.M. Cohen, 93–112. London: William Collins Sons, 1969.

———. "A Sequel to the Problem of Human Origins: The Plurality of Inhabited Worlds." In *Christianity and Evolution: Reflections on Science and Religion*, translated by René Hague, 229–36. London: William Collins Sons, 1971.

———. "Some Notes on the Mystical Sense: An Attempt at Clarification" In *Toward the Future*, translated by René Hague, 209–11. London: William Collins Sons, 1975. First published 1951.

———. "The Spirit of the Earth." In *Human Energy*, translated by J. M. Cohen, 93–112. New York: Harcourt Brace Jovanovitch, 1969. First written 1931.

———."The Stuff of the Universe (1953)." In *Activation of Energy*, translated by René Hague, 21–57. London: William Collins Sons, 1971.

———. "Universalization and Union (1942)." In Activation of Energy, translated by René Hague. London: William Collins Sons, 1971.

———. "The Vision of the Past (1942)" in The *Vision of the Past*, translated by J.M. Cohen. New York: Harper & Row, 1967.

———. "The Zest for Living." In *Activation of Energy*, translated by René Hague, 229–43. London: William Collins Sons, 1976. First written 1950.

Tiller, William. *Science and Human Transformation*. Berkeley, CA: Pavior, 1997.

Tudzynski, P., T. Correia, and U. Keller. 2001. "Biotechnology and Genetics of Ergot Alkaloids." *Applied Microbiology and Biotechnology* 57, no. 5–6 (2001): 593–605.

Vernadsky, Vladimir. *The Biosphere*, translated by D.B. Langmuir. New York: Springer-Verlag, 1998.

Von Franz, M.-L. "The Process of Individuation." In *Man and His Symbols*, edited by C. G. Jung and M.-L. Franz, 158–229. New York: Random House, 1964.

Walker, J. Samuel. *Three Mile Island: A Nuclear Crisis in Historical Perspective*. Berkeley: University of California Press, 2004.

Wallace, B. Alan. *The Art of Transforming the Mind*. Boulder, CO: Shambhala, 2022.

Ward, Benedicta, ed. *The Desert Fathers: Sayings of the Early Christian Monks*. London: Penguin Books, 2003.

Wheeler, John Archibald. *A Journey into Gravity and Spacetime*. New York: Scientific American, 1990.

Whicher, Ian. *The Integrity of the Yoga Darsana: A Reconsideration of Classical Yoga*. Albany: State University of New York Press, 1998.

Wiener, Norbert. *Cybernetics: Or Control and Communication in the Animal and the Machine*. Cambridge, MA: MIT Press, 1948.

Wilber, Ken. *One Taste: Daily Reflections on Integral Spirituality*. Boulder, CO: Shambhala, 2000.

Wiltschko, Wolfgang, and Roswitha Wiltschko. "Magnetic Orientation and Magnetoreception in Birds and Other Animals." *Journal of Comparative Physiology A: Neuroethology, Sensory, Neural, and Behavioral Physiology* 191 (2005): 675–93.

Wood, Ernest, and Paul Brunton. *Practical Yoga, Ancient and Modern: Being a New, Independent Translation of Patanjali's Yoga Aphorisms, Interpreted in the Light of Ancient and Modern Psychological Knowledge and Practical Experience*. Chatsworth, CA: Wilshire Book, 1976.

Yirka, Bob. "New Study Strengthens Olfactory Vibration-Sensing Theory." *Phys.org* website, January 29, 2013.

Zeilik, Michael, and Stephen A. Gregory. *Introductory Astronomy & Astrophysics*. 4th ed. Fort Worth, TX: Saunders College Publishing, 1998.

INDEX

Acausal Connecting Principle, 99
actual entities, 16, 260–61
agential consciousness, 6–7
alchemical ouroboros, 156
"alien hand syndrome," 86–87
Alpert, Richard, 62, 64
"Amortization of the Universe, The," (Teilhard), 144–45, 166
"Animal Magnetism" meditation, 49–50
archetypal forms, 32, 35
archetypes
    about, 33, 36, 93, 113
    collective unconscious, 118
    dimensional existence of, 117–18
    as domain of instinct psychoids, 35
    the ego and, 120–21
    as energy processes, 115–22
    as formal factors, 117
    functioning of, 113
    growth of influence, 34
    "higher," 116
    intuition of meaning of, 122
    primary role of, 115
    as psychoid factor, 115
    as psychoids, 105, 115, 117
    resonance, 115–16
awareness
    conscious, 270
    as conscious of something, 200
    deeper, 271, 273, 276
    "flow" of, 2
    inner eye, 264
    uncharted, entering, 8–9
ayahuasca, 75–78, 274

Bekenstein, Jacob D., 247–48
Bekenstein bound, 247
Bergson, Henri, 126–27, 156
bevatron collider, 20–21
Big Sur, 53, 68–69
Biosphere and Noosphere, The, (Teilhard), 154
black-body radiation curves, 220–21
Blofeld, John, 59–60
Bohm, David. *See also* holospheres
    explicit order, 16, 38, 100
    Fourier equations and, 209
    on frequency vibrations, 209
    implicate order, 17, 38, 100
    Krishnamurti dialogue, 271–72
    on motion, 34
    on nonlocality, 230
    on particles in spacetime, 223–24
    plenum and, 26
    Pribram and, 14–15, 224–25
    quantum theory and, 219

Index • 301

brain, the. See also mind
  about, 265
  cerebral hemispheres, 265–68
  holonomic imprints, 267
  left and right differences, 268
  Lilly theory of, 66
  personality and sub-personalities and, 266–67
  separate minds in, 86–89
Brain and Perception (Pribram), 216, 231

capillary system, waveguides in, 251–55
Carr, Bernard, 20, 226
centration, 144, 155–56, 159
centered universe, 196
centro-complexity
  about, 137
  direct effect of, 137
  evolution, 192
  increase in consciousness and, 183
  increasing, 134, 155, 168
  organo-psychic field of, 186
  through noosphere humanization, 168
centrogenesis, 155, 185, 186–91, 195
"Centrology" (Teilhard), 137, 139, 182–86, 187, 191–96
Chalmers, David, 115, 234
Chew, Geoffrey F., 214–15
"Christ in Matter" (Teilhard), 130–31
Christogenesis, 170
Closed Unoriented Bosonic String Theory, 24, 38
collective consciousness
  about, 94, 103
  the ego and, 119–20
  global social, 120
  ontogenesis of, 189
  ranges of, 118
  regions of, 119

collective thought, flux of, 166
collective unconscious, 12–13, 103, 118
colliders, 20–22, 26–31
Combs, Allan, ix–xi
complexes, 94, 183
complex-frequency domain, 259
Concerning Matters of Heaven and the World to Come (Fechner), 82
consciousness
  agential, 6–7
  approaches to understanding, 10
  architecture of, 155
  awakening of, 184
  centers of, 179
  change of, 122
  collective, 94, 103, 118
  "cosmic," 271
  energy model of, 151
  essence of, 41
  evolutionary arc of, 155
  exploration of, 45–46
  frequency continuum of, 107
  "hard problem" of, 115
  hidden dimensions of, 10–11
  in human body, 255
  integral, 197
  Jung assumption of, 102
  multiperspectival, 151–52
  mutation of, 197
  networks of, 41–42
  as perceiving mind centers, 86
  personal, 94
  phenomenon of, 157–58
  reflective, 188–90
  seat of, 7
  spectrum of, 107
  sub-quantum, 14–15
  superposition of, 14, 200–210
  topology of, 90–94

traditional approaches to, 157
  as waves of energy, 82
corpus callosum, 86, 87, 265, 269
cosmic consciousness, 50, 271
"Cosmic Life" (Teilhard), 146
crazy wisdom (*yeshe chölwa*), 66
Creative Evolution (Bergson), 126–27
Crowley, Aleister, 59, 60–61
cybernetics, 208–9

death, 7, 55–56, 82, 83–86, 178, 198–99
Death-Barrier, 198–99
"Death-Barrier and Co-Reflection" (Teilhard), 141
Democritus, x–xi
dimensional analysis, 256
dimensions. *See also* hidden dimensions
  breakdown of, 11
  concept, 23
  of space, 23–24, 26
  time, 24
Dirac, Paul, 215
Dirac diagram, 231
Divine Milieu, The (Teilhard), 134, 142–43
doom-scrolling, 1
"dual consciousness" theory, 86

ego, the
  about, 100–102
  archetypes and, 120–21
  in collective consciousness, 119–20
  in Jung's model, 93
  nuclear, 188
  in ocean of consciousness, 118
  peripheral, 188
  personal memory and, 106
  wavelength, 255
electromagnetic energy, 70, 174–75
electromagnetic field theory, 42

electromagnetic spectrum, 62, 106, 111
electromagnetic waves, 108–9, 238
Elements of Psychophysics (Fechner), 81–83
Eliade, Mircea, 54
energy
  archetypes and, 115–22
  axial component, 171
  controlled, 176
  electromagnetic, 70, 174–75
  holoflux, 271–72
  incorporated, 176
  love-consciousness, 177
  materialized psychic, 172
  physical, 172
  power outputs comparison, 175
  processes, archetypes as, 115–22
  of the psyche, 256
  spiritualized, 176
  tangential component, 171
eu-centric domain, 193
explicit order, 16, 38, 100, 229, 238

Fast Fourier Transform (FFT), 217
Fechner, Gustav
  about, 81–82
  on communion, 84
  Concerning Matters of Heaven and the World to Come, 82
  on consciousness, 277
  on conscious universe, 84
  Elements of Psychophysics, 81–83
  focus of work, 83
  on higher thought, 85
  Little Book of Life after Death, 80–81, 83–86
  metaphysical worldview, 87
  Nanna, or the Soul-Life of Plants, 82
  "Nightview" and "Dayview" concepts, 82
  on psychophysical movement, 85

on psychophysics, 12, 81–83
on separate minds in one brain, 86–89
"souls" and, 83
Feynman, Richard, 61, 219
fiber-optic network, 252
Fourier, Jean Baptiste, 257, 258
Fourier analysis, 208–9
Fourier transform and equations
   about, 202
   benefits of, 203
   Bohm and, 209
   common applications, 206–8
   Dirac diagram, 214–15
   inverse transform and, 205
   iPhone application, 206
   linking spacetime and hidden dimensions, 257–60
   music application, 207
   software for executing, 205
   time and frequency transformations, 204, 205–6
   use of, 204, 205
Fox, Charles R., 88–89
frequency domain, 216, 259
Freud, Sigmund, 91–92

Gebser, Jean, 197
grand unified theory (GUT), 226
Grof, Stanislav, 273
Growth of the Divine Milieu, The (Teilhard), 143–44
Gurdjieff, G. I., 218, 270

Hamilton Pool, UFO at, 56–65, 69
Hawking, Stephen, 24–25, 29–30
hidden dimensions. *See also* consciousness
   about, 11, 24–26
   author coming to exploration of, 48–78
   experiences and entities, 40
   experimental proof of, 26–31
   exploration of, 11, 15–16, 45
   first experience of, 42–47
   of mind, 264–77
   overview, 10
   psychoids in, 112
   spacetime link, 257–60
   string theory and, 29
Higgs boson particle, 8, 30, 31
higher worlds, 38–41, 45
Hilbert space, 240
Hoffman, Donald, 6–7, 8, 41
holoflux, 17, 164, 217, 224–25
holoflux theory
   about, 14–15
   Bohmian, 164
   implicate order and, 164
   multiple "big bangs" and, 234
   spirit in, 163–64
   transition zone, 164
holographic principle, 247
holomovement, 233
holopixels, 244–46
holoplenum
   about, 16
   display, 246
   energy waves from, 164
   holograms versus, 237
   holospheres in, 240
   of quantum black holes, 243–44
   spacetime projection from, 234–35
   topology of, 238–43
holospheres
   about, 16, 26
   holoplenum of, 235
   isospheres and, 196–97, 249
   moving out from, 196–97
   as origin of a flux packet, 240
   theory of, 163

Holy Trinity, 165
"Hominization" (Teilhard), 141
"Human Energy" (Teilhard), 176–78, 181–82
Human Phenomenon, The (Teilhard), 139, 146–47, 180–81
Huxley, Aldous, 62
Hynek, Allen, xi
hyperphysics. *See also* Teilhard de Chardin, Pierre
   about, 13, 123–26, 150–51
   architecture of consciousness and, 155
   emergence of, 151
   nature of theories, 124
   of noosphere, 13

I Ching, 98
implicate order
   about, 17, 38, 100
   cosmology and, 233–34
   explicit order resonance, 238
   explicit order transition, 229
   as ground beyond time, 238
   holoflux process as, 164
   quantum shells of, 242
   quantum vacuum, 241
   spectral flux and, 230–33
   unfolding into spacetime, 248–51
inner sounds, experience of, 67–71
inner space, integration of, 5
instinct psychoids, 35, 106, 113, 114, 117
instincts, 93, 110, 118
integral consciousness, 197
introspection, 38, 45, 84, 88, 275
intuition, 122, 127, 177, 238, 274
"isms," 120–21
isospheres, 196–97, 241–42, 243, 249–50

James, William, 45, 50, 80, 83–84, 87–88, 209–10

Jung, Carl. *See also* psychoids
   about, 90–91
   archetypes, 33, 93, 113–22
   butterfly symbol and, 91, 97–98
   on dimensional analysis, 256
   Freud and, 91–92
   human ego and, 100–102
   on individuation, 122
   intersection of psychic regions and, 95
   map of human consciousness, 95
   Pauli and, 98–100
   psyche and, 90, 97–98
   psyche model, 90, 92–93
   psyche-spirit-Self, 105
   on psychic processes, 110
   on spacetime continuum, 116
   "Spirit and Life," 94–97
   states of the unconscious, 101–2
   Symbols of Transformation, 92
   topology of consciousness, 90–94
   unconscious and, 35

Kahneman, Daniel, 86
Kekulé, August, 227–29
Krishnamurti, Jiddhu, 271–72
Kuo, Francis F., 204

Laminated Spacetime theory, 237
Languages of the Brain (Pribram), 213–14
Large Hadron Collider
   about, 21, 27
   construction of, 27
   Future Circular Collider and, 28
   Higgs boson collision, 30–31
   magnets, 22
   string manipulation warning, 29–30
   typical experiment within, 28
Leary, Timothy, 62, 64
Leibniz, Gottfried Wilhelm, 186

Leonard Crow Dog, 71–74
Le Roy, Pierre, 138, 139–40, 145
life and spirit, 96–97
Lilly, John, 43–44, 66–67, 273
linear ion accelerator, 52
Little Book of Life after Death, The (Fechner), 80–81, 83–86
love-consciousness energy, 177
LSD, 43, 53–55, 57, 72
Lubac, Henri de, 123, 147, 180

"Man's Place in the Universe" (Teilhard), 183–84
"Mass on the World, The" (Teilhard), 145
materialized psychic energy, 172
Maxwell, James Clerk, 82
McKenna, Terence, 54, 271
meditation, 67, 69–70, 71, 76
Mesmer, Franz Anton, 48–49
metaverse, the. *See also* Higgs boson particle
  about, 11, 17, 32
  archetypal forms and, 32–34
  holonomic, geocentric topology of, 238–43
  opening to, 272–74
microtubules, 252, 253
mind
  about, 93
  hemispherical, 264–69
  left cerebral, 268, 269
  normal, silencing, 272
  right cerebral, 268
  third, 270–72
Minkowski, Hermann, 5–6
moiré patterns, 59
Monfried, Henry and Armgart, 135
morality of balance, 160–62
morality of growth, 161

morality of movement, 160–62
Morgan, Conway Lloyd, 136–37
morphic resonance, 11, 34–36, 115
morphogenetic fields, 238
M-theory, 24, 38
multiperspectival consciousness, 151–52

Nanna, or the Soul-Life of Plants (Fechner), 82
nanobots, psychic, 75–78
"Nature of the Point Omega, The" (Teilhard), 181–82
near-death experiences, xi
neural networks, 6–7, 41
nitrous oxide, 50–51
nonduality, 14, 73, 100, 188, 197, 246
noosphere
  about, 12–13, 17, 153–54, 172–73
  approaches to, 154
  concept, 141, 154
  energetic, 13
  history of, 153–54
  hyperphysics of, 13
  incorporated, 176
  locating, 173–76
  megacentre, 189
  noogenesis and, 169, 189
  Omega point and, 168–70, 179–80
  one in many and, 178–82
  re-union, 195
  understanding of, 154
nous, 153
nuclear ego, 188

Omega point
  about, 29, 144, 260
  attributes, 190
  as "a bi-polar union," 191
  at center, 190, 191

maximum complexity, 190
noosphere and, 168–70, 179–80
as Planck holosphere, 181
radiation of, 194
as universal Center, 166
"On the Nature of the Psyche" (Jung), 98
"orchestrated objective reduction," 114
"Organization of Total Human Energy" (Teilhard), 178–80
ouroboros, 156, 227
Ouspensky, Pyotr Demianovich, 63–64, 218

Patañjali, *yoga sutras*, 71, 74–75
Pauli, Wolfgang, 98–100
Peat, F. David, 218–19
peripheral ego, 188
personal consciousness, 94
personal unconscious, 118
peyote, 62, 71–74
Phaéton, x
phase diagrams, 203
"Phenomenon of Spirituality, The" (Teilhard), 157
phyletic centricity, 188
physical energy, 172
pilot wave theory, 246
Planck, Max, 17, 110, 220–21
Planck constants, 222–23
Planck length
　about, 17
　calculation of, 222
　equation, 221
　holospheres, 24–25, 196–97
　isospheres, 241, 248
　quantum black holes, 234
　qubits, 248
　wavelengths and, 110
Planck time, 17, 223, 234–35, 236–37

plenum, 26
plus esse, 191–92
premise, this book, 4
Pribram, Karl
　about, 14–15, 213–14
　Bohm and, 14–15, 224–25
　holoflux experience, 217
　holonomic brain theory, 213–15
　hypothesis, 214
　on mantra stimulation, 218
　on "potential reality," 212
　spectral flux, 230–33
　Transcendental Meditation (TM) and, 217–18
Pribram-Bohm holoflux model, 224–25
Programming and Metaprogramming the Human Bio-Computer (Lilly), 66
psyche, the
　about, 90–94, 97–98, 103
　energy spectrum of, 256
　Jung's model of, 93–94, 98, 104
　mathematical metaphor for, 106–7
　metaphorical map, 114
　as multi-dimensioned projection, 103
　as spectrum of frequencies, 104–5, 107
　unconscious regions of, 101–2
psyche model, 93–94, 98, 104
psychic nanobots, 75–78
psychic processes, 90, 93, 107, 110, 116
psychoids. *See also* Jung, Carl
　about, 11, 12, 33, 93, 103
　archetypes as, 105, 106, 115, 117
　communication, 104
　communion with, 37
　on frequency spectrum, 36
　instinct, 35, 106, 113, 114, 117
　as operational in every dimension, 112
　resonance, 113

sizes, 113
sun as, 112
psychonauts, ix, 8–9, 18, 45–46
psychophysical movement, 85
psychophysics, 81–83, 87, 89

quantum black holes, 243–44
Quantum Brain Mechanics (QBD), 254
quantum potential, 192–94, 244, 246, 250
quantum shells, 242, 243
quantum theory, 196, 216, 219, 234
quantum vacuum, 241
qubits, 247–48

reader's guide, this book, 9–16
reflection, 184–86, 189
reflective consciousness, 188–90
religious traditions, 3–4, 120–21
resonance
  archetype, 115–16
  direct, 104, 112
  effect, 250
  electromagnetic, 42
  explicit order, 238
  frequency, 14, 42
  morphic, 11, 34–36, 115
  participatory, 77

Sanskrit language, 74–75
science, 2–3, 4
Self, 93, 270, 272–74
self-reflection, 185–86
Sheldrake, Rupert, 34, 36, 112, 115, 238
sine waves, 108–11
soul, as multidimensional hologram, 260–61
sounds, inner, 67–71
space
  axis of scales, 229
  dimensional, scales of, 228
  granularity, 219–25
  granularity of, 229–30
  isospheric shells of, 241–42
  limits of, 226–30
spacetime
  about, 5–6
  domain outside of, 231
  four-dimensional, 5–6
  as headset, 7
  laminae of, 237
  link with hidden dimensions, 257–60
  projection from the holoplenum, 234–35
  psyches within, 6–7
  reality visualization, 245
  unfolding implicate order into, 248–51
spatial complexity, 189
Spencer-Brown, G., 232–33
spirit, 94–97, 155, 156–57, 163–66
"Spirit and Life" (Jung), 94–97
Steiner, Rudolf, 39, 40, 92, 269
still point, 260
string theory
  about, 6, 24–25
  collider outcomes and, 35
  hidden dimensions and, 29
  knowledge of higher worlds and, 38–41
  M-theory, 24, 38
  string concepts and, 26
Summer of Love (1967), 53–56
sun, as psychoid, 112
superposition of consciousness, 14, 200–210
superposition principle, 200–204
Swedenborg, Emanuel, 276
Symbols of Transformation (Jung), 92
synchronicity, 98, 99–100

Teilhard de Chardin, Pierre. *See also* hyperphysics
  Bergson and, 126–27
  Biosphere and Noosphere, The, 154
  Catholic view, 170
  on centers of consciousness, 179
  centration and, 144, 155–56, 159
  centro-complexity and, 137, 169, 182–83, 186, 192
  centrogenesis and, 155, 185, 186–91, 195
  in China, 142, 145–46
  on convergence, 186
  Death-Barrier and, 198–99
  death of, 147–48
  "direct stimulation" and, 193
  The Divine Milieu, 134, 142–43
  on energy, 171–72
  in French Army, 128–31
  as geologist/paleontologist, 136
  as "The Great Monad," 139
  "heightening of antagonism," 167
  The Human Phenomenon, 139, 146–47, 180–81
  on inner being, 134
  integral life experience, 123–31
  isospheres and, 196–97
  Le Roy and, 138, 139–40
  on love, 170, 177
  moralities, 160–62
  multiperspectival consciousness, 151–52
  mystical sense, 135–48
  Omega point, 29, 144, 166, 168–70, 178–82
  The Phenomenon of Man, 124
  psychic current and, 143
  reflection and, 184–86
  science-based understanding, 124–25
  spiral, 168
  spirit and, 155, 156–57, 163–66
  studies, 126–28
  on tendencies of consciousness, 167
  Trinity, 165
  at Verdun, 129–31
  vision, 131–35
temporal complexity, 189
"the Voice of the Silence," 4
Tibetan Book of the Dead, The, 62–63
time, 219–25, 234–35, 236–37, 259
tinnitus, 68, 274
Trungpa Rinpoche, Chögyam, 65–66

UFO at Hamilton Pool, 56–65
unconscious
  collective, 12–13, 103, 118
  Jung and, 35, 101–2
  personal, 118
  states of, 101–2
universe, the, 5, 32, 37, 196. *See also* metaverse, the

Vallée, Jacques, xi
Vernadsky, Vladimir Ivanovich, 140, 145
Void, the, 273

Wallace, Alan B., 32–34, 35
waveguides, 251–55
Weiser's Bookstore, 65
Wheeler, John Archibald, 247, 248
Whicher, Ian, 74
Whitehead, Alfred North, 260–61
Wholeness and the Implicate Order (Bohm), 233, 246
Wiener, Norbert, 208–9, 258–59
Wilber, Ken, 271, 273
Wunt, Wilhelm, 275

Yoga Sutras, 71, 74–75